AF341608

THE SEARCH FOR ANTIVIRAL DRUGS

Case Histories
from Concept to Clinic

THE SEARCH FOR ANTIVIRAL DRUGS

Case Histories from Concept to Clinic

Julian Adams
Vincent J. Merluzzi

Editors

96 Illustrations

Birkhäuser
Boston · Basel · Berlin

Julian Adams
Boehringer Ingelheim
  Pharmaceuticals, Inc.
900 Ridgebury Road
P.O. Box 368
Ridgefield, CT 06877-0368
USA

Vincent J. Merluzzi
Boehringer Ingelheim
  Pharmaceuticals, Inc.
900 Ridgebury Road
P.O. Box 368
Ridgefield, CT 06877-0368
USA

Library of Congress Cataloging-in-Publication Data
The Search for antiviral drugs : case histories from concept to clinic
  / Julian Adams, Vincent J. Merluzzi, editors.
       p. cm.
    Includes bibliographical references and index.
    ISBN 0-8176-3606-4 (acid-free paper : Boston). — ISBN
  3-7643-3606-4 (acid-free paper : Berlin)
    1. Antiviral agents. 2. Nucleosides—Derivatives—Therapeutic
  use. 3. HIV infections—Chemotherapy. I. Adams, Julian, 1954–   .
  II. Merluzzi, Vincent J., 1949–   .
  RM411.S39 1993
  616.9′250651—dc20   93-16405
                                                                        CIP

Printed on acid-free paper.

          Birkhäuser 

ISBN 0-8176-3606-4
ISBN 3-7643-3606-4

Typeset by Lind Graphics, Inc., Upper Saddle River, NJ.
Printed and bound by Edwards Brothers, Inc., Ann Arbor, MI.
Printed in the United States of America.

9 8 7 6 5 4 3 2 1

Contents

Preface

Perspectives for Future Antiviral Drugs

The field of antiviral drug discovery is certainly making enormous advances and continues to progress at a remarkable rate. We have witnessed that the early days of antivirals produced nucleoside analogues that demonstrated landmark achievements for the treatment of viral infections. Currently, second- and third-generation nucleoside analogues are offering more specific and safer alternatives to the initially discovered drugs. A great deal of effort has been expended in studying the mechanism of action and metabolism of nucleoside antivirals, and this has allowed a detailed understanding at the molecular and cellular level as to how these drugs work. Nevertheless, in contrast to the wide array of agents available for the treatment of bacterial infections, the number of antiviral drugs remains relatively small. The reasons for this stem from the very nature of the role of nucleosides as genetic building blocks for eukaryotic and prokaryotic replication. Host cell replicative machinery makes use of the very same nucleoside triphosphate substrates for DNA and RNA synthesis as do many of the nucleoside-based antivirals. Furthermore, we have also learned that the nucleoside analogues are generally administered in the unphosphorylated form and must then be converted by either the host cell or viral kinases. This, too, can pose a problem since different cell types can phosphorylate at different rates and at different times during the cell cycle. Thus, the particular pharmacokinetic and pharmacodynamic properties as well as the specificity vis-à-vis cellular DNA and RNA polymerases necessarily complicate matters for the development of new nucleoside analogues. In general, nucleoside antivirals have exhibited a less than optimal therapeutic index.

An additional parameter that must be considered in the field of antiviral drugs is the issue of organism resistance. Viral genomic mutation through selective pressure induced by drug therapy is a serious limitation in devising antiviral strategies. The more effective the drug, the more rapidly the emergence of an escape mutant, through random error driven viral replication, leads to decreased sensitivity to the antiviral drug. Both DNA and RNA viral polymerases and reverse transcriptase are notoriously poor at maintaining genomic fidelity. Mismatched base pairing occurs with higher frequency than in mammalian polymerases. Furthermore, the complex editing functions and DNA repair are not operative in primitive viral organisms. Thus, it follows that for all known antiviral therapeutics, resistance (or reduced sensitivity) to drug eventually emerges upon prolonged treatment. This is discussed in the following chapters in detail for individual therapies.

How are we then to overcome the limitations of reduced efficacy of an antiviral drug as a result of phenotypic viral resistance? Two conceivable strategies come to mind:

The first involves the possible treatment of viral infections using combinations of drugs that may in an additive or synergistic manner further block viral replication to an infinitesimally slow rate, thus allowing the host's immune system to clear the infection. Since mutation is replication dependent, total blockade of replication should theoretically arrest viral spread and increase the time to the emergence of resistant viral strains. This is obviously an oversimplification since issues such as viral latency and reactivation are not addressed, but it still offers a potential solution to viral spreading. Combination therapy is successfully used in the treatment of bacterial infections or with aggressive cancers.

The second strategy that may be envisioned is to attack some aspect of the viral replicative cycle whereby, if mutation occurs at the site of drug–protein receptor interaction, the virus cannot replicate since the specific residues of the protein are fundamental to the replicative cycle and by definition may not be altered. Such approaches are alluded to in this book, for example, where the antiviral is an enzyme inhibitor interacting with active site catalytic residues, or where the drug may be interacting with cellular cofactors. This second strategy remains theoretical but may in fact be the most desirable approach.

One of the most important advances that has contributed to the discovery of new antiviral drugs is the scientific contribution of molecular biology in helping to dissect and unravel the viral life cycle. Molecular virology has led to a sophisticated understanding of the structure and function of viral proteins and enzymes, and this has uncovered many attractive targets for drug discovery. Not only has fundamental scientific knowledge increased regarding the pathophysiology of viruses, but also practical techniques in molecular cloning and structural biology have made possible the production of viral target proteins and enzymes as laboratory tools in order to screen for new inhibitors. In addition, techniques in biomolecular structure such as X-ray crystallography and nuclear magnetic resonance (NMR) have provided three-dimensional models to better design drugs that are more specific and bind with higher affinities. Such technological advances have made mechanistically based drug design a practical reality in recent years and provide a rational basis to target human pathogenic viruses without compromising the host's normal physiology.

In the final two chapters, two promising unique approaches toward antiviral therapy for the treatment of rhinoviral infection (by Marlin) and herpes infection (by Liuzzi and Déziel) are described. While both of these projects are still in preclinical stages of maturity and as such depart from the main theme of this book, they are being highlighted in these final chapters to demonstrate the creativity and opportunity in rational drug design and indicate the future trends for discovery of antiviral therapeutic agents.

We hope that you will agree that the chapters presented in this book offer a true narrative of the discovery and development process. Both successes and failures in experimentation, theory, and decision making are brought to bear. Success or failure in the clinical setting is not the major issue but rather the concept and process of getting a lead molecule to the clinical stage. Chapter 1 offers a historical perspective of antiviral development, followed by chapters highlighting current antiviral compounds that have been successful preclinical candidates and are worthy of clinical research. We hope that you find the contents of this book interesting from several points of view, not the least of which are how concepts are started and how decision-making events have led to clinical trials.

Contributors

Julian Adams Boehringer Ingelheim Pharmaceuticals, Inc., 900 Ridgebury Road, P.O. Box 368, Ridgefield, Connecticut 06877, USA

Koen Andries Department of Virology, Janssen Research Foundation, Turnhoutseweg 30, B-2340 Beerse, Belgium

Robert Déziel Bio-Méga, Inc., 2100 rue Cunard, Laval, Québec, Canada H7S 2G5

Mark E. Goldman Ligand Pharmaceuticals, 9393 Towne Centre Drive, San Diego, California 92121, USA

Ming-Chu Hsu Departments of Virology and Anti-Infective Chemistry, Hoffmann–La Roche, Nutley, New Jersey 07110, USA

Herbert E. Kaufman LSU Eye Center, 2020 Gravier Street, Suite B, New Orleans, Louisiana 70112, USA

Michel Liuzzi Bio-Méga, Inc., 2100 rue Cunard, Laval, Québec, Canada H7S 2G5

Steven D. Marlin Boehringer Ingelheim Pharmaceuticals, Inc., 900 Ridgebury Road, P.O. Box 368, Ridgefield, Connecticut 06877, USA

Vincent J. Merluzzi Boehringer Ingelheim Pharmaceuticals, Inc., 900 Ridgebury Road, P.O. Box 368, Ridgefield, Connecticut 06877, USA

Kathryn H. Pattishall Burroughs Wellcome Co., 3030 Cornwallis Road, Research Triangle Park, North Carolina 27709, USA

Rudi Pauwels Rega Institute for Medical Research, Minderbroedersstraat 10, B-3000 Leuven, Belgium

Sally Redshaw Roche Products, Ltd., P.O. Box 8, Broadwater Road, Welwyn Garden City, Hertfordshire AL7 3AY, England

Noel A. Roberts Roche Products, Ltd., P.O. Box 8, Broadwater Road, Welwyn Garden City, Hertfordshire AL7 3AY, England

Steve Tam Departments of Virology and Anti-Infective Chemistry, Hoffmann–La Roche, Nutley, New Jersey 07110, USA

THE SEARCH FOR ANTIVIRAL DRUGS
Case Histories from Concept to Clinic

1

Introduction:
The First Effective Antiviral

HERBERT E. KAUFMAN

Introduction

During the 1950s there was an explosion of research on antivirals by a number of superb scientists with excellent resources. However, the lack of progress in discovering a clinically useful agent led to widespread discouragement. In fact, the company to which I brought idoxuridine—the first clinically effective antiviral—had closed its antiviral research section a week before my arrival. At that time, to my knowledge, no pharmaceutical company was engaged in antiviral research.

The discovery of idoxuridine as a treatment for the most common infection of the cornea, herpes simplex keratitis, not only provided a therapeutic approach to a clinically important disease but also renewed hope that clinically effective virus chemotherapy was possible. The widespread acceptance of such a drug, however, depended on far more than the discovery of the agent and an understanding of how to administer it. The new drug would never have had the impact that it had without a clear understanding of the kinds of clinical disease that came after herpes infection and the pathogenesis of the various phases of disease. For example, disease due to hypersensitivity and previous tissue damage was not treatable by antiviral agents alone. Suddenly, clinicians required an understanding of the mechanism of different phases of infection, in addition to a therapeutically useful agent, in order to treat disease effectively.

At the time all of this occurred, I was a resident in ophthalmology at the Massachusetts Eye and Ear Infirmary in Boston. At my disposal, I had a small laboratory in a decaying, condemned building, one technician, and no specific funds for the study of viruses. The discovery of

The Search for Antiviral Drugs
Julian Adams and Vincent J. Merluzzi, Editors
© 1993 Birkhäuser Boston

idoxuridine as an effective antiviral treatment for herpetic disease, the subsequent elucidation of the pathophysiology of types of disease not caused by rapidly multiplying virus, and the education of mature clinicians who had no previous reason to consider differences in pathogenesis posed a variety of challenges for someone still early in his career. In this introductory chapter I will try to provide some feeling for what this discovery process was like.

Background

Before my entrance into the field of virus chemotherapy, there had been substantial research and diverse approaches to the subject. Although this will not be an extensive review, some feeling for what came before is important in elucidating the discovery process.

One popular approach to the treatment of virus disease, especially disease that could be treated topically, was a search for agents that would inactivate extracellular virus. It was known that denaturing agents, protoplasmic poisons, detergents, lipid solvents, and many other chemicals, when mixed with cell-free virus, could destroy the virus. On that basis, Sery and Furgiuele (1961) evaluated a large series of chemicals, especially heavy metals, for this purpose. Also tested were a series of polysaccharides, and even antibody, which might inactivate virus before infection but seemed to have little effect once infection was established.

The first step in the discovery process was the assumption that these earlier approaches would not be clinically useful and that the search for a clinically effective therapeutic agent for virus disease would have to begin elsewhere. I assumed that for an antiviral agent to be clinically useful it would need to be effective after virus had entered the cell. Looking for ideas, I assembled the relevant literature but found it terribly confusing.

Tissue culture studies had yielded a large and bewildering array of antimetabolites that prevented or reduced virus multiplication in vitro when administered before or even simultaneously with infection. The earliest compounds were analogues of tryptophan and methionine (Rasmussen et al., 1951; Mathews and Smith, 1955), but amino acid analogues such as fluorophenylalanine and β-phenylserine (Dickinson and Thompson, 1957; Zimmermann and Schafer, 1960; Scholtissek and Rott, 1961; Brown et al., 1961; Pons and Preston, 1961) also received considerable attention. Although these compounds could dramatically reduce the amount of virus produced by infected tissue, they seemed to me to be relatively nonspecific inhibitors of protein synthesis and, despite all of the beautiful studies and imaginative experiments, I did not believe that this approach could yield selectivity or therapeutic utility.

Similarly, the general inhibition of energy metabolism was found to reduce viral production and infectivity. Not only the reduction of oxygen tension but also agents that interfered with oxidative metabolism diminished viral multiplication in parallel with the decrease in aerobic respiration. Fluoroacetate poisoned the Krebs cycle, and dinitrophenol uncoupled high-energy phosphate bonds; other agents such as malonate, iodoacetate, and *p*-chloromercuribenzoate were also used. Again, however, I decided arbitrarily that these agents, which reduced virus multiplication in tissue culture, were unlikely to be clinically useful. It seemed evident that virus synthesis in the infected cell would require large amounts of energy for the construction of virus DNA and proteins. There was no evidence, however, that any of these pathways were induced in the cell by the virus.

To be sure, poisoning the cell in a nonspecific way early in the course of virus infection slowed infection and/or reduced virus yield, but none of these agents aborted virus disease in vivo. In fact, the plethora of early tissue culture studies showing in vitro activity that could not be correlated with in vivo utility both gave rise to a general distrust of tissue culture systems for evaluating antiviral compounds and created an atmosphere of pessimism about the likelihood of finding useful agents that would be therapeutically active in animals or humans.

One of the giants in the early work with virus chemotherapy was Igor Tamm at the Rockefeller Institute in New York. Horsfall and Tamm, in a review (1957, p. 339), summarized their view of the field as follows:

Only a few of the virus diseases of man can be prevented effectively and there is no means for the adequate treatment of any, excluding those caused by the psittacosis–lymphogranuloma group of virus-like agents. Thus, at present, the control of human virus diseases is commonly not feasible and most are accepted as inevitable ailments which run their course without therapeutic modification. This places virus diseases in a uniquely unsatisfactory position, for with no other category of infectious processes is management so underdeveloped and the need for it so large.

Tamm had worked extensively with the benzimidazole compounds, only to be frustrated by the tantalizing hint of some tissue culture activity but no therapeutic usefulness. Despite these frustrations, when I went to talk with Tamm and his student, Hans Eggers, about the possibility of a different approach to the treatment of virus disease, they urged me to continue. Even though I was an unknown first-year resident with minimal formal training in virology, they received me warmly and were encouraging about the value of my proceeding in this field, in which they and others had expended so much fruitless effort. Without this encouragement, I certainly never would have embarked on the studies outlined below.

The Hypothesis

Many studies had been done on purine and pyrimidine antagonists in tissue culture. Some of these agents even protected animals to a modest extent when given before infection or at about the same time as infection, but none had been therapeutically useful in animals or humans. I assumed that if an agent was to be therapeutically useful it would have to act on a specific virus-encoded metabolic site. Although many antimetabolites could, in a nonspecific way, poison cellular metabolism and inhibit virus infection in tissue culture, none of these agents halted established infection, even in tissue culture, and none of the drugs was therapeutically useful in established virus infections in animals. In the case of purine and pyrimidine antagonists, the virus did not appear to initiate new ways of synthesizing purine and pyrimidine nucleotides; since the synthesis of these compounds was by normal cellular pathways, it was probably not amenable to selective inhibition. In bacteriophage, new pathways of nucleotide metabolism are required for the synthesis of hydroxymethylcytosine, but these specific pathways are not present in animal cells.

In order to synthesize new virus from cellular products, it seemed that both a specific DNA primer and a new polymerase system were probably required. It appeared likely that the affinities of either the viral-encoded enzyme or the primer for the nucleotide might well be different from those of the standard cellular polymerase and that an agent which acted at the point of assembly of the nucleotides (at the DNA polymerase step) would offer some hope for relatively specific antiviral effects. If this hypothesis was correct, agents that interfered with nucleotide synthesis at the earlier stage could not, with any specificity, inhibit virus multiplication. Furthermore, the rapid rate of DNA synthesis after virus infection suggested to me that the affinity of the virus assembly enzymes for nucleotides must be higher than the affinity of the normal cellular polymerase system, i.e., the virus probably had "first call" on any available nucleotide pool. In the early 1960s, there was little available support for this hypothesis, but since then considerable corroborative evidence has been discovered. We know that specific DNA polymerase is synthesized in herpes-infected cells. Additionally, when thymidine synthesis and DNA synthesis are totally inhibited by fluorodeoxyuridine and small amounts of thymidine are added to the virus-infected culture, virus synthesis continues even when insufficient nucleotide is present to permit cellular DNA production (Easterbrook and Davern, 1963).

With this hypothesis in mind, I looked through the biochemical literature for an agent that might inhibit or act at the DNA polymerase step, although at the time there was no way to judge relative selectivity. In the late 1950s, William Prusoff, a professor of pharmacology at Yale, had synthesized 5-iodo-2'-deoxyuridine (IDU; IDUR; idoxuridine)

(Prusoff, 1959), which was subsequently studied as an oncolytic agent (Calabresi et al., 1961; Welch, 1961; Calabresi and Welch, 1962; Calabresi, 1963). Delamore and Prusoff (1962) showed that although the major site of inhibition varied from cell to cell and although kinase inhibition did occur in some cells such as Ehrlich's ascites tumor cells, IDU appeared primarily to inhibit DNA polymerase and to be incorporated into an ineffective DNA. This suggested to me, for the first time, a possible drug which could act at a site that might permit inhibition and selective incorporation through a specific virus-induced enzyme. Additionally, although the significance was not apparent at the time, Herrmann (1961) had shown that IDU could inhibit established virus infection using an in vitro system in which a sheet of cells was infected with virus and overlaid with agar, and antiviral agents were applied via disk diffusion.

Testing the Hypothesis

Prusoff's brilliant biochemical pharmacology had provided a compound that not only fit my hypothesis but also could be purchased from a chemical supply house. The problem was that I had no specific funds with which to purchase either the drug or the animals needed to test it. However, my assistant, Emily (Maloney) Varnell, and I were running a small laboratory in which we did studies on ocular inflammation, and the hospital allowed us to charge for doing toxoplasma dye tests on infected humans and animals as an aid to their treating physicians. The funds earned from this testing were used to purchase the first lot of IDU and the rabbits for the first in vivo trials.

All of this work would have been infinitely easier and more efficient if I had had the courage to introduce myself and my ideas to Bill Prusoff, who had synthesized IDU. At the time, as a beginning resident, however, I was reluctant to approach him and did not avail myself of his generosity and vast fund of knowledge.

For example, when we attempted to formulate usable eye drops, we were frustrated to find that the drug was almost insoluble and that the highest concentration we could obtain was 0.1%. Realizing that most of a drop applied to the eye is washed out and that less than one-tenth of 1% of the available drug penetrates the cornea, we wondered whether topical drops could deliver a therapeutically effective dose. However, on the principle of "nothing ventured, nothing gained," we elected to proceed.

The model we chose was different from what had been used before. In the past, rabbits were infected and a general grading system was used similar to the Draize score (Draize et al., 1944), which included evaluations of redness of the conjunctiva, redness of the lids, inflammation of the inside of the eye, cloudiness of the cornea, as well as the size of the ulcer itself. I decided that, for our purposes, this system was imprecise and misleading. We had good evidence that the branching, tree-shaped dendritic epithelial ulcer caused by herpes simplex was the result of cellular destruction by rapidly multiplying virus. Thus, the size of the ulcer seemed to be a more appropriate indicator of infection, and I elected to ignore the adjunct epiphenomena. In our system, the ulcers were graded numerically on a masked basis: 1+ for an ulcer covering 25% of the total area of the cornea; 2+ for 50%; 3+ for 75%; and 4+ for a totally ulcerated cornea (Kaufman and Maloney, 1962). This turned out to be a wise decision; the new grading system was sufficiently precise and reliable so that not only could therapeutic activity be detected but also dose–response curves of different drugs could be obtained, ED_{50}'s could be computed, and relative drug potencies could be accurately compared (Figure 1) (Kaufman, 1965).

Because the concentration of IDU obtainable in the drops was so low and because I thought that IDU would be a competitive antagonist of thymidine uptake, I decided that for therapy the drug should be given

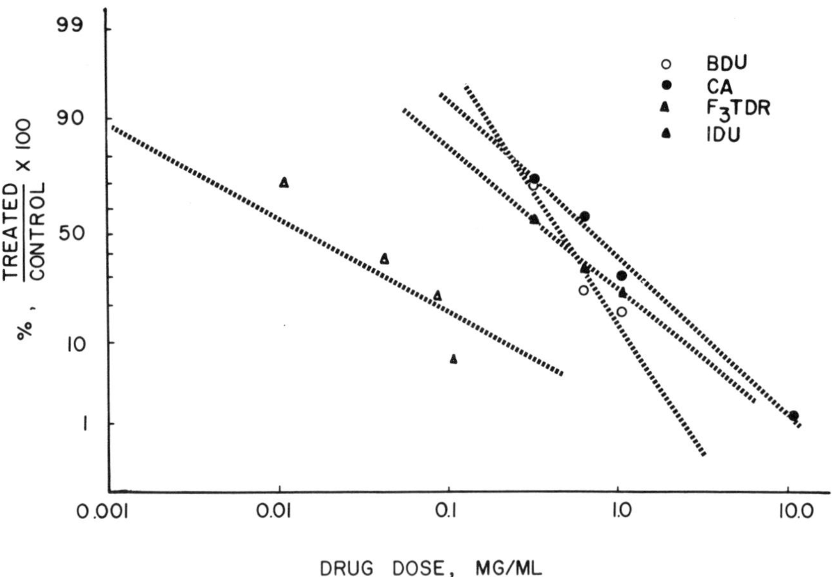

FIGURE 1. The concentration of antivirals in eye drops is plotted against the extent of the corneal ulcer after 4 days of therapy in the rabbit model of herpes keratitis. The dose–response curve that results permits a calculation of ED_{50} and the comparison of drug potency. It also facilitates the study of additive or synergistic drug effects. *Key:* BDU, 5-bromo-2'-deoxyuridine; CA, cytosine arabinoside; F_3TDR, trifluorothymidine; IDU, 5-iodo-2'-deoxyuridine. Reprinted with permission from Kaufman HE (1965): *In vivo* studies with antiviral agents. *Ann NY Acad Sci* 130:168–180.

every hour during the day and every 2 hours at night. This decision also turned out to be a good one. Even later when the drug was tested in human patients, this arbitrarily chosen therapeutic regimen proved to be the most effective and continued to be used in clinical practice.

Since we had no money to pay for help with these experiments, three of us—my assistant, Mrs. Varnell, a bright, young, creative beginning ophthalmology resident, Dr. Anthony Nesburn, and I—undertook to do the study ourselves. Each night, one of us slept on a cot in the center of an incredibly tiny, pungent animal room, setting an alarm and getting up every 2 hours to treat the animals. Of course, this was strictly a personal endeavor, and we were still obliged to carry a normal work load the following day (Figures 2 and 3).

In controlled studies of IDU drops administered at various times after infection, there was no question that the treated rabbits got better relatively quickly compared to the controls and that in these animals IDU was an extremely effective therapeutic agent (Kaufman, 1962; Kaufman et al., 1962a). Following these studies, Dr. Claes Dohlman and others at the Massachusetts Eye and Ear Infirmary permitted me to see and treat a relatively large number of patients with herpetic dendritic keratitis, and again the therapeutic effect was striking (Kaufman et al.,

FIGURE 2. This decrepit, condemned building in downtown Boston on the campus of the Massachusetts Eye and Ear Infirmary housed our laboratory. Our space consisted of three tiny rooms on the second floor, formerly the apartment of the superintendent of the hospital cleaning staff. There was no elevator in the building, and all animals and equipment had to be carried up the steps.

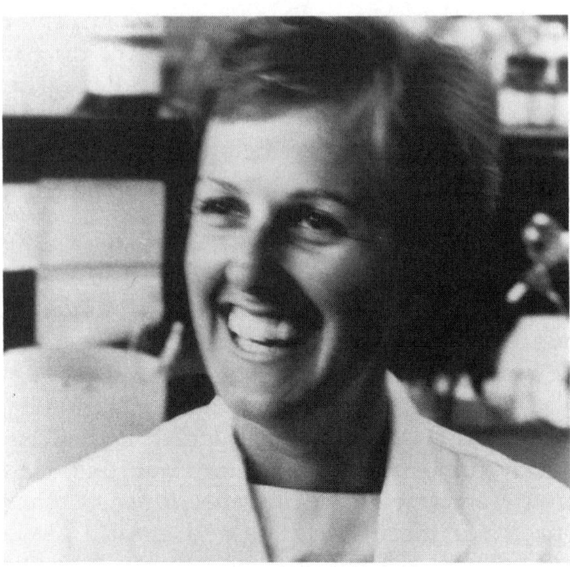

Figure 3. Mrs. Emily Varnell is a chemical engineer who not only did expert research but was willing to sleep on a portable cot in a pungent animal room, waking up every 2 hours at night to treat the animals, for which she received no overtime pay or compensatory time off.

1962b). Although my original clinical studies were uncontrolled, they were rapidly followed by five controlled double-blind human studies, all of which demonstrated significant therapeutic activity for IDU (Paterson et al., 1963; Burns, 1963; Laibson and Leopold, 1964; Jepson, 1964; Hart et al., 1965).

Since it was clear from the activity of IDU that the iodine group made the 5-iodo compound an analogue of thymidine, and since the van der Waals radii of the bromo- and chloro-substituted 5-uridines were only slightly smaller than that of IDU, we tested them and found them also to be active (Kaufman et al., 1962c; Kaufman, 1963a). They seemed more toxic than the original compound, however, and were never used clinically. Substances that inhibited DNA synthesis but did not act on the final polymerase step, such as 5-fluorodeoxyuridine, were not clinically useful as therapeutic antivirals—a finding which further confirmed my original hypothesis that active agents would have to exert their effect at the DNA polymerase step.

Potential purine and pyrimidine analogues seemed to fall into two broad groups. All analogues might bind either to the DNA polymerase or to the primer, which would confer some direct inhibition of synthesis. Since the helical backbone of DNA is a chain of glucose molecules linked by 3′,5′-phosphodiester bonds from which the bases hang like the teeth on a key, the active substituted bases with normal sugars would permit synthesis of the sugar backbone of the DNA but result in an abnormal DNA code that could not specify functioning, multiplying virus. The analogues with normal bases but abnormal sugars, such as the arabinosides and later acyclovir, would terminate synthesis of the sugar backbone and function as chain terminators. Other activities might also be present; for example, 5-trifluoromethylthymidine is an important inhibitor of thymidylic synthetase, but this and virtually all of the active compounds appeared to act primarily on the polymerization process (Figure 4).

The hypothesis—that to be effective, purine and pyrimidine analogues must work on the final polymerization step of viral DNA—was formulated before any drug testing was done. The selection of IDU, therefore, was not the result of a random screening process, but rather a specific choice based on a specific hypothesis. Even so, that the first

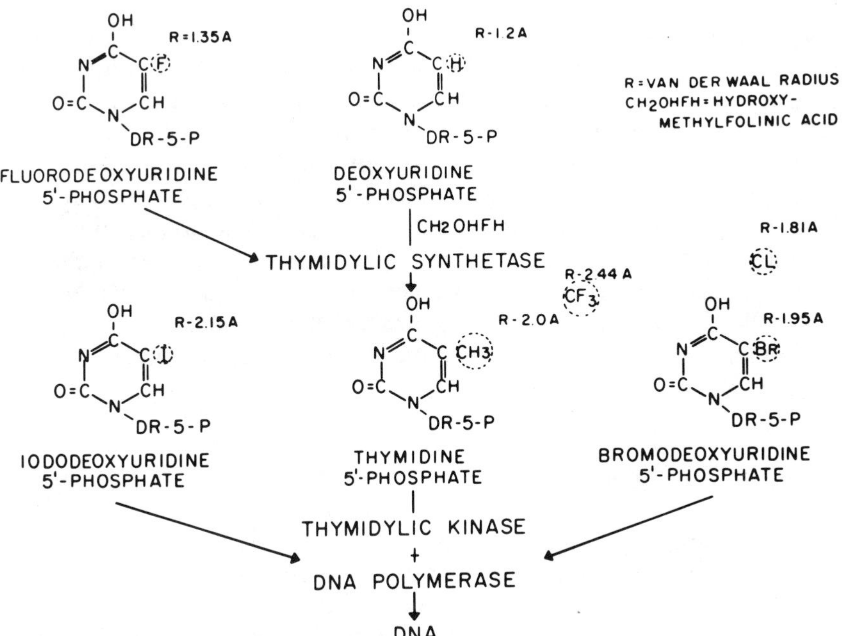

FIGURE 4. Effective antivirals like idoxuridine interact directly with DNA polymerase. The van der Waals radius numbers *(R)* are similar to that of thymidine. Compounds such as 5-fluorodeoxyuridine, which inhibited earlier steps in synthesis, were not active. Reprinted with permission from Kaufman HE, Maloney ED, Nesburn AB (1962): Comparison of specific antiviral agents in herpes simplex keratitis. *Invest Ophthalmol* 1:686–692.

drug we tested was effective was both somewhat fortuitous and extremely fortunate, because we certainly didn't have enough money to test any significant number of additional compounds.

Problems with the Clinical Use of the First Antiviral Drug

Herpes simplex keratitis accounts for approximately 500,000 infections of the cornea in the United States each year; it is the most common specific corneal infection in humans, and occurs all over the world. One might suppose that the introduction of the first therapeutically useful antiviral, idoxuridine (Stoxil™, Smith Kline & French, Philadelphia, PA; Herplex™, Allergan, Irvine, CA), which was clearly effective in laboratory animals and in human patients, would be greeted with uniform acceptance and relief, but considerable controversy and disappointment remained.

The basic scientists involved in virology research recognized the importance of this discovery and, as a consequence, I was invited to present my results at the prestigious Perspectives in Virology meeting in New York City in February 1962. Clinicians, however, remained skeptical because not all herpes patients appeared to respond. This was because the antiviral was most useful only for the epithelial dendritic keratitis caused by multiplying virus. Other kinds of postinfectious keratitis were severe and blinding, and yet were not amenable to antiviral chemotherapy. For virus chemotherapy to reach its true potential in clinical practice, the pathogenesis and treatment of these manifestations of herpes needed to be rapidly elucidated and accepted by the profession.

Dendritic Keratitis

Dendritic keratitis was known to be caused by rapidly multiplying virus and was found to be amenable to antiviral chemotherapy. Even though an effective antiviral agent might limit epithelial cell division to some extent, the epithelium of the cornea is approximately six layers thick and defects can heal by sliding, so that the integrity of the superficial layers can be reestablished without much cell division. Thus, the epithelium was an ideal substrate for treatment with antimetabolites, even those that are not totally selective.

Sometimes the dendritic ulcer, which began as the branching tree type (Figure 5), expanded to cause denudation of a significant portion of the cornea and appeared as a "geographic ulcer"—an ulcer in the shape

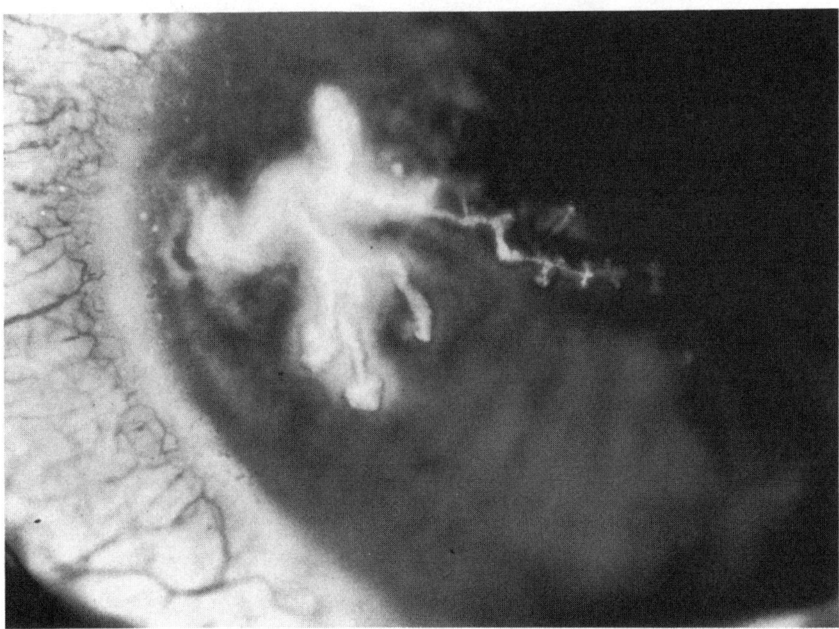

FIGURE 5. An example of dendritic keratitis, the common form of HSV-1 infection of the cornea, caused by multiplying virus and treatable with antiviral drugs.

of a map. Geographic ulcers were also amenable to virus chemotherapy, although of course the larger lesions healed more slowly.

Disciform Keratitis

In some patients, a swelling and clouding of the cornea accompanied by pain and loss of vision were observed to follow an epithelial herpes infection (Figure 6). This problem, termed *disciform edema* or *disciform keratitis*, was clearly not due to virus multiplying rapidly in the tissue. When specimens were taken from clinical cases, multiplying virus generally was not found. The condition often persisted for at least a period of months, causing significant morbidity, and if left unchecked caused permanent scarring and vascularization of the cornea.

There was good reason to think that this clinical disease was caused by a hypersensitivity reaction to the virus and its products (probably a glycoprotein) and that it could be treated with topical corticosteroids. It was known, for example, that a significant number of patients vaccinated for small pox did not develop skin lesions, even though they developed antibody. Further studies indicated that the skin lesions of vaccinia, and perhaps those of most exanthems, are largely hypersen-

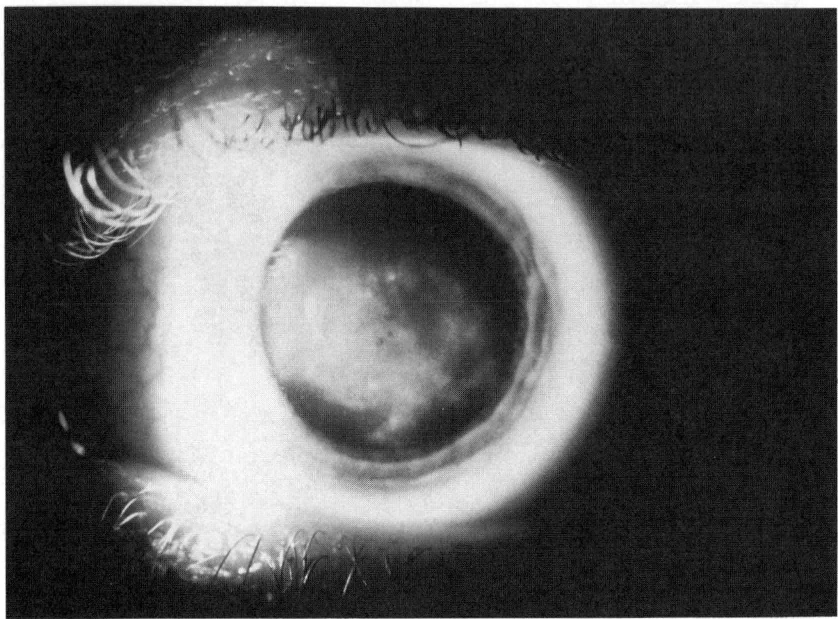

FIGURE 6. Disciform keratitis is primarily a hypersensitivity reaction to viral antigen and is treated with a combination of corticosteroids and an antiviral.

sitivity reactions that occur after virus multiplication has reached its peak rather than as a result of specific cytodestruction by the virus. In the eye, it appeared that the same phenomenon might be at work.

However, treatment of the herpes-infected eye with corticosteroids frequently caused a rapid exacerbation of epithelial disease and severe ulceration of the cornea. Treating disciform keratitis with corticosteroids, therefore, posed a significant risk of permanent corneal damage. Because corticosteroids offered considerable potential therapeutic benefit in the treatment of the blinding inflammatory symptoms of disciform keratitis although the risks for causing recurrent epithelial disease were substantial, I did a series of studies in rabbits and found that corticosteroids and antivirals had opposing activities on epithelial disease (Kaufman and Maloney, 1962). If a large dose of corticosteroids was used alone, the disease became significantly worse; however, if a large dose of antiviral was applied simultaneously, the antiviral could counterbalance and neutralize the exacerbative corticosteroid activity in a competitive way. Based on these studies, I offered the clinician the first useful approach to this disease—the combined use of corticosteroids and antivirals to treat disciform keratitis (Kaufman, 1963b; Kaufman et al., 1963). It was clear that the corticosteroids quieted the disease and that the antiviral, while not therapeutically important, made the corticosteroids safe. Of course, the corticosteroids didn't really cure anything, although the benefits were striking; what they did was restore vision, relieve symptoms, and make the patient comfortable and clinically well until the condition ran its course. We found that most patients could be taken off the corticosteroid drops after about 3 months without fear of recurrence.

Necrotizing Keratitis

A smaller number of patients developed white necrotic lesions in the corneal stroma (Figure 7), which appeared to be due to a combination of virus multiplication and hypersensitivity. In these patients, topical antivirals were of questionable value in eliminating the multiplying virus because of its location deep in the cornea. (Even with today's more potent drugs, antiviral treatment, topical or systemic, is of uncertain value.) The hypersensitivity component, however, could be treated with corticosteroids and antiviral combined and, although the results were less predictable than they were in patients with disciform keratitis, the treatment was valuable.

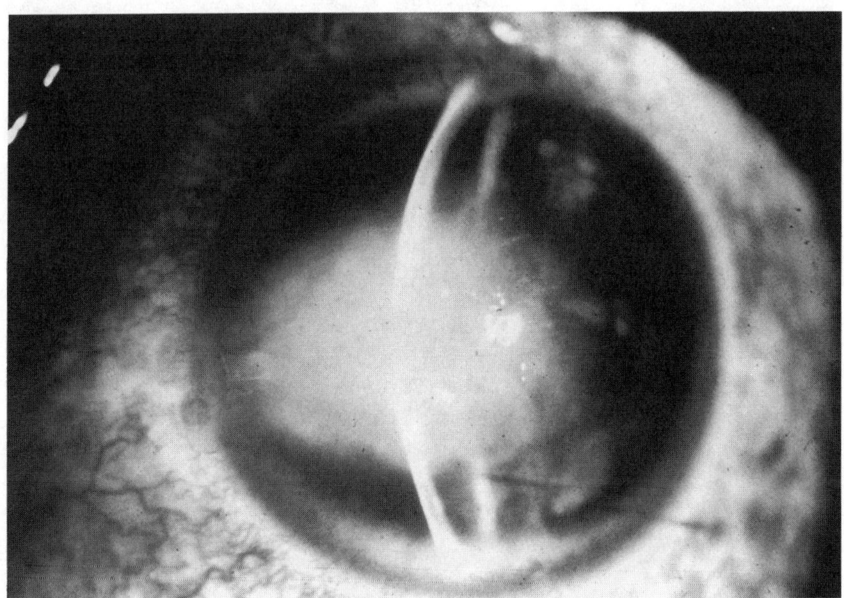

FIGURE 7. In necrotizing keratitis, a deep stromal hypersensitivity reaction to viral products is combined with stromal virus multiplication, which makes treatment very difficult.

Metaherpetic Keratitis: Postinfectious Ulcers

The final syndrome that required treatment was a recurrent corneal ulceration that developed after the original herpes infection had apparently been eliminated (Figure 8). Our work indicated that this kind of ulceration was similar to ulcers that occurred after mechanical damage to the cornea of other sorts, i.e., any nonspecific corneal trauma or infection. We suggested that the pathogenesis of these "recurrent erosions" was not recurring virus and that this condition must be differentiated from true recurrences of virus disease.

In this disorder, called *metaherpetic keratitis,* the postinfectious ulcers appeared to be caused by damage to basement membranes and hemidesmosomes (Figure 9) that did not permit the superficial epithelium to adhere properly to the stroma. In these cases there might or might not be complete healing after infection, but even though no virus was present there were recurrent epithelial ulcerations (Kaufman, 1964). The treatment of this kind of disease—which had previously been confused with active herpes—with antivirals and frequent drops was harmful

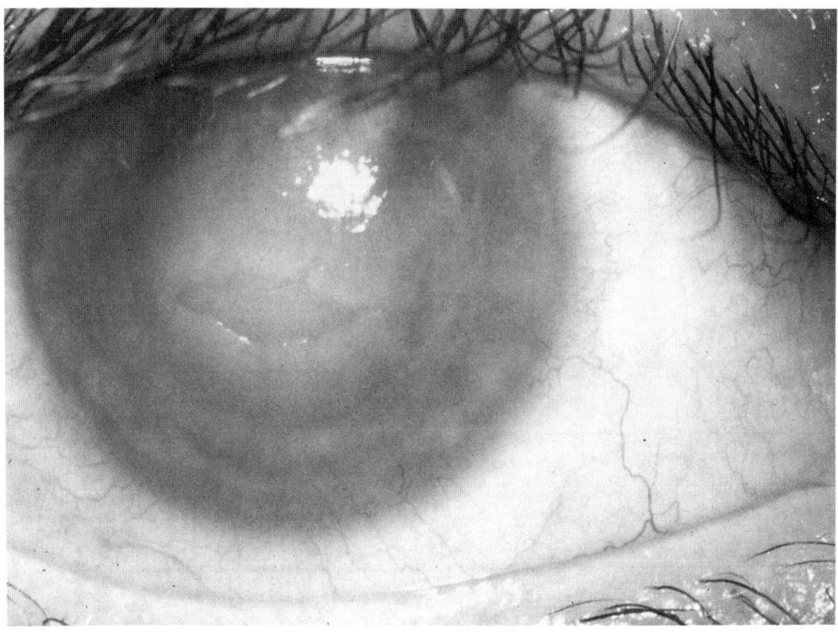

FIGURE 8. Metaherpetic keratitis is a recurrent ulceration that develops after HSV-1 infection but is caused by nonspecific damage to basement membrane and hemidesmosomes, and not by multiplying virus. Therefore, treatment is protective, including ointments and a bandage soft contact lens to prevent irritation from the movement of the eyelids.

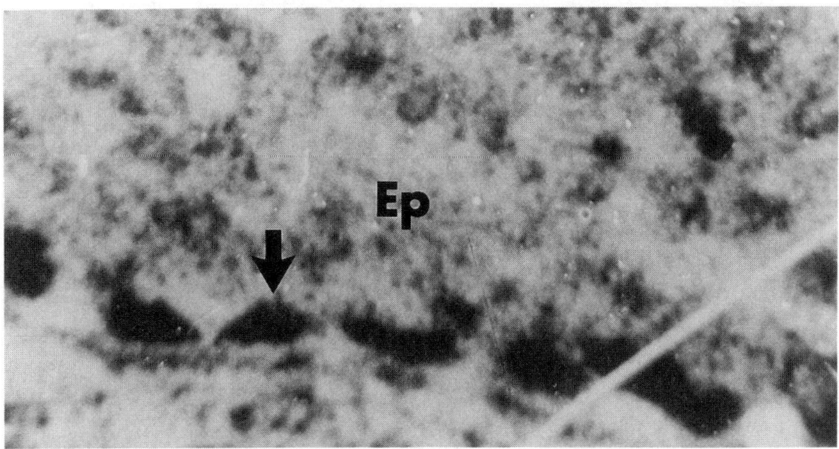

FIGURE 9. These epithelial attachment sites, the hemidesmosomes (arrow), reform if the newly healing epithelium (Ep) is encouraged to migrate and is kept in place. Reprinted with permission from Kaufman HE (1978): Herpetic keratitis. *Invest Ophthalmol Vis Sci* 17:941–957.

because the constant opening and closing of the eyelids rubbed up and down over the epithelium that was trying to migrate over and attach to damaged basement membrane; also, the administration of any medication that might be even slightly irritating or toxic was contraindicated. Identifying this syndrome and differentiating it from that of true recurrences of virus disease permitted a different kind of therapeutic treatment: the application of bland ointment and a patch, or even the use of a soft contact lens, to protect the surface of the cornea from blinking lids and to allow the epithelium to migrate over and attach to the damaged basement membrane, finally reforming normal hemidesmosomes.

The clarification of the pathogenesis of these various syndromes—and of our developing treatments for them as well as for the initial virus infection—was important for the clinical use of the first antiviral and made virus chemotherapy an accepted part of clinical practice.

Virus Strain Specificity of Herpetic Disease Characteristics

Another problem, although a much smaller one, was that it became clinically apparent that some infections were far more virulent than others and some responded better to treatment. Centifanto, Wander, and I did much of the original work to establish the fact that the herpes genome encoded by different HSV-1 strains produced very different properties in terms of virulence (Wander et al., 1980; Centifanto-Fitzgerald and Kaufman, 1981; Centifanto-Fitzgerald et al., 1982a,b,c; Smeraglia et al., 1982). In fact, working with Roizman at the University of Chicago, we made and studied recombinants of known strains with specific infection patterns (Centifanto-Fitzgerald et al., 1982a). We demonstrated that even the shape of the dendritic ulcer on the cornea of the eye was determined at a specific locus on the HSV genome and was relatively independent of the host and size of infection. In these experiments, epithelial disease and stromal disease were shown to segregate separately on the viral genome. These early experiments which showed that infection and response depended on a diversity of viral genetic material and not just on host resistance explained some of the differences not only of clinical infection in humans but also in animal experiments from laboratory to laboratory.

Modern Ocular Virus Chemotherapy

The most important additional step in the modern treatment of virus disease was the introduction by Kaufman and Heidelberger (1964) of trifluorothymidine (trifluridine; Viroptic™; Burroughs Wellcome, Research Triangle Park, NC) as a topical antiviral (Figure 10). This agent could be given as an eye drop only five times a day with no administration necessary at night and was more effective both as a therapeutic agent and a corticosteroid antagonist than IDU. In fact, with trifluridine, 97% of patients with dendritic herpes were healed within 2 weeks (Wellings et al., 1972). Trifluridine is the preferred medication for herpes keratitis in the United States today.

In 1977, the introduction of acyclovir (Zovirax™, Burroughs Wellcome, Research Triangle Park, NC) offered a new possibility for the treatment of ocular herpes (Elion et al., 1977; Schaeffer et al., 1978). However, topical acyclovir is not likely to supplant trifluridine as the drug of choice. Although the toxicities of the two drugs are, surprisingly, not significantly different in acute studies and the two drugs seem to have comparable antiviral activity when given topically, acyclovir is not generally used as a topical ocular agent in the United States because of the potential for viral resistance. All of the purine and pyrimidine antagonists must be phosphorylated before they can interact with DNA polymerase, and acyclovir is selectively phosphorylated by viral thymidine kinase. Thus virus can become resistant to acyclovir by three mechanisms: (1) it can drop out its thymidine kinase and still multiply (although these thymidine kinase–deficient strains seem less virulent,

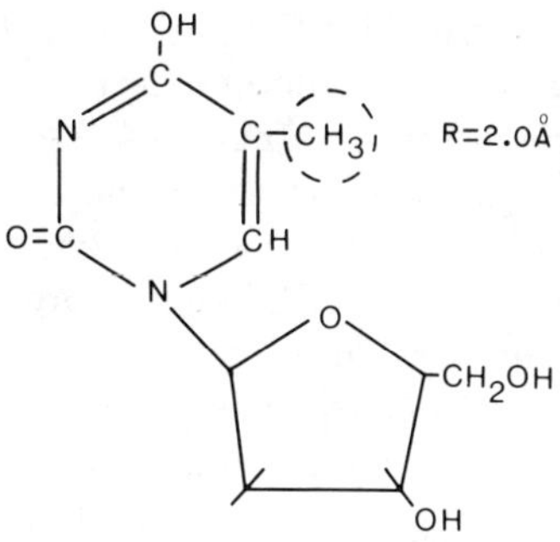

THYMIDINE

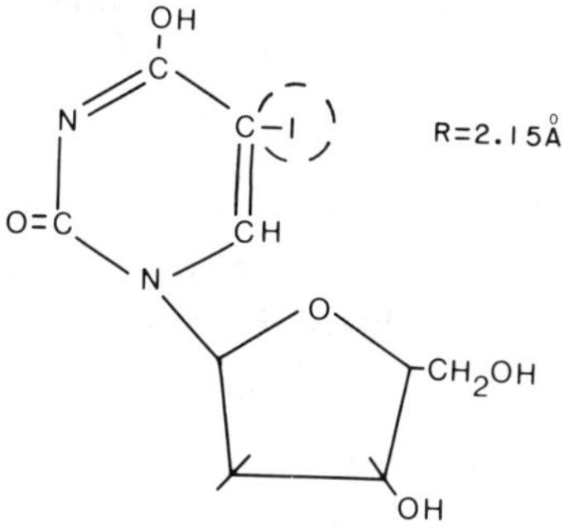

R=2.0Å

IODODEOXYURIDINE   R=2.15Å

TRIFLUOROTHYMIDINE   R=2.44Å

FIGURE 10. Structures of the antiviral drugs idoxuridine (iododeoxyuridine) and trifluorothymidine. Note the similarity to the normal cellular metabolite thymidine. Adapted and redrawn with permission from Kaufman HE (1978): Herpetic keratitis. *Invest Ophthalmol Vis Sci* 17:941–957.

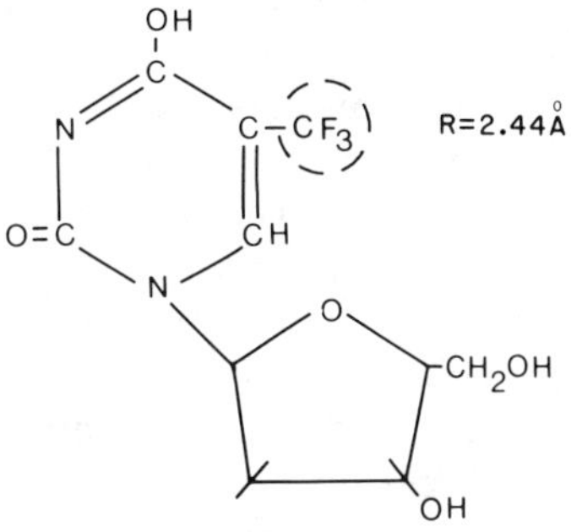

they do produce disease in humans); (2) it can alter its thymidine kinase; and (3) it can alter its DNA polymerase. If, through widespread topical use, a pool of acyclovir-resistant virus were created, there is a real risk that patients with systemic disease could not be treated effectively with this and other thymidine kinase–selective antivirals. Additionally, acyclovir is not totally selective and can alter host DNA, thus producing fetal abnormalities.

Orally administered acyclovir is secreted in the tears, and it can be used to treat dendritic keratitis in special cases, for instance, children and in disabled adults who cannot easily manage eye drops. So far, systemically administered acyclovir has not been shown to be of benefit in stromal herpetic disease, even for necrotizing stromal keratitis, where a multiplying virus seems to play a part. A multicenter, controlled study is being done by the National Eye Institute to determine whether oral acyclovir is of benefit in treating stromal keratitis or iritis, but to date there is no conclusive evidence in favor of this.

Oral acyclovir has also not been shown to prevent recurrences of ocular disease. There is no question that patients with recurring genital lesions and patients prone to cutaneous herpes after bone marrow transplantation or immunosuppression can markedly reduce their risk of recurrent herpes by taking oral acyclovir. Studies in the eye, however, do not show that oral acyclovir (Sanitato et al., 1984) or even the potent bromovinyldeoxyuridine (BVDU) prevents recurrences (Kaufman et al., 1983). I believe that the difference lies in the amount of virus multiplication needed to cause disease in the end organ. In the case of genital disease, virus traveling down the nerve from the ganglion is not sufficient to produce a genital or cutaneous lesion, but rather significant local viral multiplication must take place. This, I believe, is what is prevented by oral acyclovir. In contrast, the corneal lesions in the eye are so small that virus traveling down the nerve from the ganglion may be enough to produce a lesion without significant local viral multiplication and so acyclovir is ineffective.

A new class of compounds has been developed that may be even more useful for preventing recurrences. These compounds selectively inhibit only viral thymidine kinase, so they would not be active in peripheral tissues where there is enough viral thymidine kinase from the cell to permit virus multiplication. However, when I heard about them I realized that the brain and ganglia contain little or no thymidine kinase, since there is little or no cell division or DNA synthesis in the central nervous system, and so these compounds might somehow inhibit the virus in the ganglion and prevent reactivation. A preliminary study of one of these compounds (5'-ethynylthymidine) indicates that it does reduce the risk of recurrences in the eyes of primates previously infected and subject to spontaneous recurrences (Kaufman et al., 1991).

Other more active and soluble viral thymidine kinase inhibitors may ultimately be of clinical value. Similarly, Öberg and colleagues (1982) have found that the phosphonates act directly to inhibit DNA polymerase and do not require phosphorylation, and they too are an important additional class of active compounds against herpes and other viruses.

Summary

Idoxuridine was recognized as the first clinically effective antiviral. It effectively treated the most common corneal infection of humans, herpes keratitis, which was associated with significant morbidity. Its proper clinical use required not only an understanding of the antiviral agent but also an understanding of the different manifestations of herpes in the eye, such as those caused by hypersensitivity or previous tissue damage, and the development of methods to treat these manifestations as well as the acute viral infection. When IDU was found, I was a resident, and when the pathogenesis of the different manifestations was elucidated and I recommended treatments such as the combined use of corticosteroids and antivirals to safely treat disease due to hypersensitivity, such as disciform keratitis, or the treatments to encourage healing in corneas previously damaged (metaherpetic keratitis), I was at the very beginning of my career. Yet it was imperative that others in my field differentiate these syndromes which had never required differentiation before, and rationally use these new types of therapy. Writing, lecturing around the country, and teaching whenever I could accomplished what was needed. Although these principles are now commonly used and well established, at the time of their introduction there was a degree of disbelief and controversy that is hard to imagine now.

Acknowledgment

This work was supported in part by US Public Health grant EY02672 from the National Eye Institute, National Institutes of Health, Bethesda, Maryland.

References

Brown F, Planterose DN, Stewart DL (1961): Effect of *p*-fluorophenylalanine on the multiplication of foot and mouth disease virus. *Nature (London)* 191:414–415

Burns RP (1963): A double-blind study of IDU in human herpes simplex keratitis. *Arch Ophthalmol* 70:381–384

Calabresi P (1963): Current status of clinical investigations with 6-azauridine, 5-iodo-2′-deoxyuridine and related derivatives. *Cancer Res* 23:1260–1267

Calabresi P, Welch AD (1962): Chemotherapy of neoplastic diseases. *Annu Rev Med* 13:147–202

Calabresi P, Cardoso SS, Finch SC, Kligerman MM, von Essen CF, Chu MY, Welch AD (1961): Initial clinical studies with 5-iodo-2′-deoxyuridine. *Cancer Res* 21:550–559

Centifanto-Fitzgerald Y, Kaufman HE (1981): Herpes simplex virus strain specificity and ocular disease. In *Herpetische Augenerkrankungen*, Sundmacher R, ed. Munich: JF Bergmann Verlag, pp 19–24

Centifanto-Fitzgerald YM, Yamaguchi T, Kaufman HE, Tognon M, Roizman B (1982a): Ocular disease pattern induced by herpes simplex virus is genetically determined by a specific region of viral DNA. *J Exp Med* 155:475–489

Centifanto-Fitzgerald YM, Varnell ED, Kaufman HE (1982b): Initial herpes simplex virus type 1 infection prevents ganglionic superinfection by other strains. *Infect Immun* 35:1125–1132

Centifanto-Fitzgerald YM, Fenger T, Kaufman HE (1982c): Virus proteins in herpetic keratitis. *Exp Eye Res* 35:425–441

Delamore IW, Prusoff WH (1962): Effect of 5-iodo-2′-deoxyuridine on the biosynthesis of phosphorylated derivatives of thymidine. *Biochem Pharmacol* 11:101–112

Dickinson L, Thompson MJ (1957): The antiviral action of threo-β-phenylserine. *Br J Pharmacol* 12:66–73

Draize JN, Woodard G, Cavery HO (1944): Methods for the study of irritation and toxicity of substances applied topically to the skin and mucous membranes. *J Pharmacol Exp Ther* 82:377–390

Easterbrook KB, Davern CI (1963): The effects of 5-bromodeoxyuridine on the multiplication of vaccinia virus. *Virology* 19:509–520

Elion GB, Furman PA, Fyfe JA, De Miranda P, Beauchamp L, Schaeffer HJ (1977): Selectivity of action of an antiherpetic agent, 9-(2-hydroxyethoxy-methyl)guanine. *Proc Natl Acad Sci USA* 74:5716–5720

Hart DRL, Brightman VJF, Readshaw GG, Porter GTJ, Tully MJ (1965): Treatment of human herpes simplex keratitis with idoxuridine. *Arch Ophthalmol* 73:623–634

Herrmann EC Jr (1961): Plaque inhibition test for detection of specific inhibitors of DNA containing viruses. *Proc Soc Exp Biol Med* 107:142–145

Horsfall FL Jr, Tamm I (1957): Chemotherapy of viral and rickettsial diseases. *Annu Rev Microbiol* 11:339–370

Jepson CN (1964): Treatment of herpes simplex of the cornea with IDU. *Am J Ophthalmol* 57:213–217

Kaufman HE (1962): Clinical cure of herpes simplex keratitis by 5-iodo-2′-deoxyuridine. *Proc Soc Exp Biol Med* 109:251–252

Kaufman HE (1963a): Treatment of herpes simplex and vaccinia keratitis by 5-iodo- and 5-bromo-2′-deoxyuridine. *Perspect Virol* 111:90–107

Kaufman HE (1963b): Treatment of deep herpetic keratitis with IDU and corticosteroids. *EENT Dig* 25:37–40

Kaufman HE (1964): Epithelial erosion syndrome: Metaherpetic keratitis. *Am J Ophthalmol* 57:983–987

Kaufman HE (1965): *In vivo* studies with antiviral agents. *Ann NY Acad Sci* 130:168–180

Kaufman HE (1978): Herpetic keratitis. *Invest Ophthalmol Vis Sci* 17:941–957

Kaufman HE, Heidelberger C (1964): Therapeutic antiviral action of 5-trifluoro-methyl-2′-deoxyuridine. *Science* 145:585–586

Kaufman HE, Maloney ED (1962): IDU and hydrocortisone in experimental herpes simplex keratitis. *Arch Ophthalmol* 68:396–398

Kaufman HE, Nesburn AB, Maloney ED (1962a): IDU therapy of herpes simplex. *Arch Ophthalmol* 67:583–591

Kaufman HE, Martola E-L, Dohlman CH (1962b): The use of 5-iodo-2′-deoxyuridine (IDU) in the treatment of herpes simplex keratitis. *Arch Ophthalmol* 68:235–239

Kaufman HE, Maloney ED, Nesburn AB (1962c): Comparison of specific antiviral agents in herpes simplex keratitis. *Invest Ophthalmol* 1:686–692

Kaufman HE, Martola E-L, Dohlman CH (1963): Herpes simplex treatment with IDU and corticosteroids. *Arch Ophthalmol* 69:468–472

Kaufman HE, Varnell ED, Centifanto-Fitzgerald YM, De Clercq E, Kissling GE (1983): Oral antiviral drugs in experimental herpes simplex keratitis. *Antimicrob Agents Chemother* 24:888–891

Kaufman HE, Varnell ED, Cheng YC, Bobek M, Thompson HW, Dutschman GE (1991): Suppression of ocular herpes recurrences by a thymidine kinase inhibitor in squirrel monkeys. *Antiviral Res* 16:227–232

Laibson PR, Leopold IH (1964): An evaluation of double blind IDU therapy in 100 cases of herpetic keratitis. *Trans Am Acad Ophthalmol* 68:22

Mathews REF, Smith JD (1955): The chemotherapy of viruses. *Adv Virus Res* 3:49–148

Öberg B, Alenius S, Eriksson, B, Helgstrand E, Lundberg C, Lundström J (1982): Preclinical evaluation of the herpesvirus inhibitor foscarnet sodium. In *Herpesvirus: Clinical, Pharmacological, and Basic Aspects*, Shiota H, Cheng Y-C, Prusoff WH, eds. Amsterdam: Excerpta Medica, pp 175–178

Paterson A, Fox AD, Davies G, Maguire C, Sellers PWH, Wright P, Rice NSC, Cobb B, Jones BR (1963): Controlled studies of IDU in the treatment of herpetic keratitis. *Trans Ophthalmol Soc UK* 83:583

Pons MW, Preston WS (1961): The in vivo inhibition by β-phenylserine of rabies, myxoma and vaccinia viruses. *Virology* 15:164–172

Prusoff WH (1959): Synthesis and biological activities of iododeoxyuridine, an analog of thymidine. *Biochim Biophys Acta* 32:295–296

Rasmussen AF Jr, Clark PF, Smith SC, Elvehjem CA (1951): Effect of 6-methyltryptophan on Lansing poliomyelitis in mice. *Bacteriol Proc* pp. 93–94

Sanitato JJ, Asbell PA, Varnell ED, Kissling G, Kaufman HE (1984): Acyclovir in the treatment of herpetic stromal disease. *Am J Ophthalmol* 98:537–547

Schaeffer HJ, Beauchamp L, de Miranda P, Elion GB, Bauer DJ, Collins P (1978): 9-(2-Hydroxyethoxymethyl)guanine activity against viruses of the herpes group. *Nature (London)* 272:583–585

Scholtissek C, Rott R (1961): Influence of *p*-fluorophenylalanine on the production of viral ribonucleic acid and on the utilization of viral protein during multiplication of fowl plague virus. *Nature (London)* 191:1023–1024

Sery TW, Furgiuele FP (1961): The inactivation of herpes simplex virus by chemical agents. *Am J Ophthalmol* 51:42–57

Smeraglia R, Hochadel J, Varnell ED, Kaufman HE, Centifanto-Fitzgerald YM (1982): The role of herpes simplex virus secreted glycoproteins in herpetic keratitis. *Exp Eye Res* 35:443–459

Wander AH, Centifanto YM, Kaufman HE (1980): Strain specificity of clinical isolates of herpes simplex virus. *Arch Ophthalmol* 98:1458–1461

Welch AD (1961): Some metabolic approaches to cancer chemotherapy. *Cancer Res* 21:1475–1490

Wellings PC, Awdry PN, Bors FH, Jones BR, Brown DC, Kaufman HE (1972): Clinical evaluation of trifluorothymidine in the treatment of herpes simplex corneal ulcers. *Am J Ophthalmol* 73:932–942

Zimmermann T, Schafer W (1960): Effect of *p*-fluorophenylalanine on fowl plague virus multiplication. *Virology* 11:676–698

2

Discovery and Development of Zidovudine as the Cornerstone of Therapy to Control Human Immunodeficiency Virus Infection

KATHRYN H. PATTISHALL

Background of Antiviral Research at Wellcome

From the time Dr. George Hitchings began the research program at Wellcome Research Laboratories in 1942 to search for antagonists of nucleic acid bases, viruses were among the potential chemotherapeutic targets. A number of 5-substituted uracil derivatives and 2,6-diamino-purine, which had been identified as inhibitors of bacterial nucleic acid synthesis in *Lactobacillus casei*, were found to also interfere in tissue culture with the multiplication of vaccinia virus, a DNA virus. Simultaneously, the laboratory had discovered a new antiparasitic drug, pyrimethamine, a pyrimidine, to treat malaria and toxoplasmosis, and a new antileukemic drug, 6-mercaptopurine, in the purine series. As a result, extensive efforts were devoted to bringing these to clinical use. Unfortunately, the antiviral agents had to take a back seat, particularly because the toxicity of 2,6-diaminopurine to bone marrow discouraged further pursuit of this lead at levels required for antiviral activity. It was over ten years later that the deoxyribosides of 5-iodouracil, 5-chloro-uracil, and 5-trifluoromethyluracil were synthesized by others and found to have activity against the herpes viruses. While two of these deoxynucleosides (5-iodo- and 5-trifluoromethyluracil) were clinically useful topically for herpes keratitis, they were too toxic to be employed systemically.

Research on nucleosides at Wellcome was concentrated during the 1950s and 1960s on purine analogues in the search for new anticancer drugs. The discovery of the antiviral activity of adenine arabinoside reawakened interest in searching for antiviral agents among the nucleosides. This initiative stimulated a close collaboration between the

The Search for Antiviral Drugs
Julian Adams and Vincent J. Merluzzi, Editors
© 1993 Birkhäuser Boston

Wellcome Research Laboratories in the United States and at Beckenham in the United Kingdom, where antiviral testing was conducted. The first successes came with the finding that 2,6-diaminopurine arabinoside and guanine arabinoside had good activity against the herpes simplex viruses and vaccinia virus, in vivo as well as in vitro. There was, however, no firm commitment to pursuing the search for antiviral nucleosides at that time.

Whereas the early studies in other laboratories had been confined to 2′-deoxyribosides of 5-substituted pyrimidines, the 1970s saw a flurry of synthetic activity at Wellcome in the modification of the structure of the sugar moiety. One of the most exciting of these modifications was the substitution of an acyclic side chain for the deoxyribose. In particular, 9-[(2-hydroxyethoxy)methyl]guanine, or acyclovir, synthesized by Schaeffer and his colleagues created great excitement because of its high selectivity for the herpes viruses and its low human host toxicity. This led to a major effort at Wellcome not only to elucidate the mechanism of action and selectivity of acyclovir but also to expand the facilities and staff for all antiviral research. This research brought into sharp focus the exploitable differences between virally induced enzymes and normal cellular enzymes and made it clear that selective interference with viral nucleic acid synthesis was an achievable goal. With the launching of acyclovir (Zovirax®) in the early 1980s as a successful chemotherapeutic agent for the treatment of a variety of herpesvirus infections, the highly trained group of chemists, biologists, and clinicians at Wellcome began to turn attention to other viruses and to other nucleoside analogues. The human retroviruses were emerging as a particular challenge as medical pathogens, and the time was ripe since a major epidemic was pending.

Discovery of the Antiviral Zidovudine and Its Early Evaluation

Chemistry

In her research department at Wellcome Research Laboratories, Dr. Gertrude B. Elion established a multidisciplinary team approach to antiviral drug development.

The 3′-azido-2′,3′-deoxyribonucleosides and 2′-azido-2′-deoxyribonucleosides were targeted (Krenitsky et al., 1981), and two series of novel compounds resulted. The azide moiety was chemically reduced to the corresponding amine. Peptide derivatives attached through the nucleoside amino functionality were also synthesized; the 2′-series paled in light of what ensued with the 3′-series. Early in 1981 two 3′-azido-3′-deoxythymidines were prepared at Wellcome and entered into compound screening. Compound 22U81 was the *threo*-analogue with the 3′-azido substituent above the plane of the sugar ring, and compound 509U81 (later named zidovudine; also known as azidothymidine or AZT) was the *erythro*-analogue with the 3-azido below the plane of the sugar ring where thymidine would have a hydroxyl group (Figure 1). Both compounds were synthesized by published procedures (Lin and Prusoff, 1978; Glinski et al., 1973) and were tested at Wellcome Research Laboratories in the United States and the United Kingdom. Although zidovudine had initially been synthesized by Dr. Jerome Horwitz and associates in 1964 at the Michigan Cancer Foundation as a potential anticancer agent, studies with the compound were abandoned shortly thereafter because of a lack of activity against animal cancers.

Early Evaluation for Microbiological Activity

Both compounds were active in microbiological screens, but the minimum inhibitory concentrations (MICs) obtained with bacterial strains sensitive to zidovudine were less than 10- to greater than 100-fold lower than those obtained for 22U81. Extensive in vitro studies of zidovudine showed a limited spectrum of activity, with inhibition against a variety of gram-negative enteric bacteria, but the gram-positive bacteria were naturally resistant. The MICs in vitro were in the range of 0.1–4 g/mL for *Escherichia coli* B, *Salmonella typhimurium*, *Shigella flexneri*, *Klebsiella pneumoniae*, and *Enterobacter aerogenes*, but gram-positive bacteria such as *Streptococcus pyogenes* and *Pseudomonas aeruginosa*, as well as anaerobic bacteria, mycobacteria, and various fungi, were not inhibited. Some of these organisms found to be naturally resistant have low levels of thymidine kinase or lack it. In fact, cultures of *Escherichia* and *Salmonella* could grow out from the inhibitory effects of zidovudine with time, with the resistant mutants demonstrating much reduced levels of thymidine kinase. In addition, *Escherichia* grown in the presence of zidovudine was found to contain the mono-, di-, and triphosphates of the drug. Purified

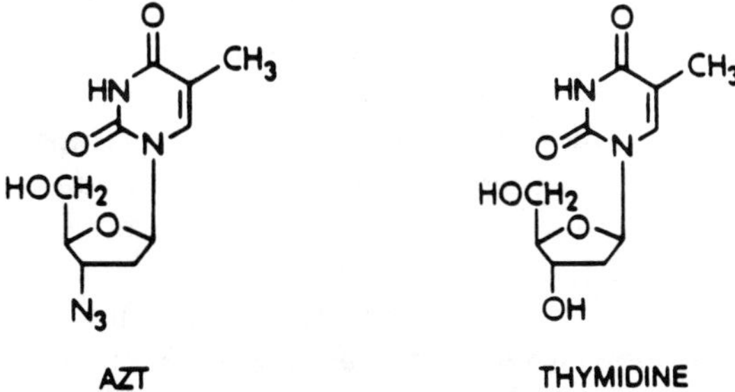

FIGURE 1. Zidovudine (AZT) and thymidine.

thymidine kinase from *Escherichia* converts the nucleoside to its monophosphate with the same efficiency as thymidine. Zidovudine binds to the enzyme and is an alternative substrate, and zidovudine triphosphate is a DNA chain terminator in the in vitro DNA polymerization catalyzed by the Klenow fragment of *E. coli* DNA polymerase 1 (Elwell et al., 1987). Therefore, incorporation into DNA II and chain termination could explain the bactericidal action of zidovudine.

In vivo, antibacterial studies with zidovudine at the US facilities and at Wellcome, Berkhamsted, in the United Kingdom demonstrated activity in experimental models. Mice were protected from life-threatening septicemic infections of *Escherichia* and acute ascending pyelonephritis. Veterinary models showed that calves were protected from fatal infections of *Salmonella dublin.* Other experimental models utilized chickens and weaning pigs. No toxicity to the compound was noted during these short-term experiments.

Other Bioactivity Assessments

Zidovudine, based upon its microbiological activity and lack of toxicity, was further assessed. Inactivity against a wide variety of DNA and RNA viruses was demonstrated: herpes simplex virus types 1 and 2, varicella virus, adenovirus type 5, influenza A, respiratory syncytial virus, rhinovirus B, yellow fever virus, measles virus, coronavirus, and bovine rotavirus. Zidovudine was also inactive against vaccina virus, vesicular stomatitis virus, and the murine leukemia, L1210, in vitro. Ultimately, tests showed it was inactive against human cytomegalovirus. Some activity is reported for zidovudine against Epstein–Barr virus (EBV), for the drug inhibits viral growth by 50% in vitro (ED_{50}) at a concentration of 3 M.

Discovery of Antiretroviral Activity

Sandra Nusinoff-Lehrman, M.D., a senior clinical research scientist, David W. Barry, M.D., the director of clinical investigation, Phil Furman, Ph.D., a senior virologist, and Marty St. Clair, a junior virologist, formed a task force early in 1984 to review the literature and to determine what approaches Wellcome should take to most quickly and efficiently direct their research efforts since the human immunodeficiency virus (HIV), a retrovirus, had been identified as the causative agent of AIDS. In October 1984, the newly discovered retrovirus HIV, then termed HTLV III or LAV, was the subject of a series of seminars given at Wellcome Research Laboratories in the United States by Drs. Françoise Barre-Sinoussi, Robert Gallo, and Samuel Broder. These discussions served as a catalyst to ensure Wellcome's involvement in the discovery and development of therapies for the new disease syndrome devastating thousands of patients, mainly in the prime of their lives. An early decision was to further develop a plaque reduction assay already in place since 1980 for an animal retrovirus, which then would be used to screen potential anti-HIV compounds.

Friend leukemia virus (F-MuLV, a murine retrovirus), Harvey sarcoma virus (HaSV, another murine retrovirus), and FG-10 (murine) cells were acquired from Kent Weinhold at Duke University. Since both F-MuLV and HaSV form plaques in FG-10 cells, the basis for the initial Wellcome assay was that if a compound inhibits the ability of the virus to grow, the number of plaques formed will be reduced. Once the murine plaque reduction assay was functioning, known antiviral compounds, such as Wellcome's acyclovir, were screened. When no exciting activities were observed with these initial candidates, the focus shifted to the dideoxynucleoside group of compounds, such as dideoxycytosine, dideoxyadenosine, dideoxyguanosine, and dideoxythymidine. While activities in the 10–20 M range were demonstrated with these compounds in the assay, the Wellcome scientists were optimistic that a compound could be identified with even greater anti-HIV activity.

Wellcome's senior organic chemists supplied 10–20 compounds representative of their synthetic expertise and established drug development programs for testing in the new plaque reduction assay. Among 12 such compounds furnished on November 2, 1984, was zidovudine. Not knowing what excitement would befall the laboratory, the compounds were analyzed at the next available opportunity.

When counting the plaques from that assay, it became obvious that none of the 18 plates with zidovudine had plaques. None at all! The virus used was apparently completely incapable of replicating in any of the concentrations tested. Notwithstanding the modest activity seen in previous assays with other compounds, no plaque formation was a rare event. The virologist, incredibly excited but also somewhat apprehensive, postulated that inoculation of those plates might not have occurred. The result was discussed by the task force team, but there were no plans to share the news widely until there was a chance to repeat the assay. But little was it realized how interested the company scientists were in the budding HIV program, for rumors began circulating almost immediately, causing feverish activity.

In quickly developing the plaque reduction assay, Wellcome virologists were fully aware that the surrogate viruses they had chosen for their retrovirus screen at this stage of the program were just that, surrogate viruses. Arrangements to have active compounds analyzed in other relevant screens, such as feline leukemia virus and HIV (known then as HTLV III), had already been made and numerous coded compounds had been sent to William Hardy (at Sloan-Kettering Cancer Center) for feline leukemia virus testing, and Sam Broder (National Cancer Institute), Kent Weinhold (Duke University), and Gerald

Quinnan (U.S. Food and Drug Administration) for analysis with HIV. The field of HIV laboratory research, very young at that time, lacked standardization, and few groups possessed the expertise sufficient to provide Wellcome with the testing required to evaluate its active compounds. As a result, several scientists using different assays were sought to confirm the antiviral activity seen and better quantitate it.

The initial indication of antiviral activity with zidovudine in each of these assays caused continued excitement. After discussions with the highest management, the largest mobilization of the human and monetary resources necessary to bring a drug to patients in the most efficient and timely manner occurred in Wellcome's history and has set the standard for the development of any new drug for the treatment of a life-threatening disease.

Medicinal Chemistry

Many of the studies described above required the continual support from chemists who furnished the compound initially and who developed methods for making the labeled compound necessary for mechanistic studies. Analogues of zidovudine, as well as compounds where the 3'-azido moiety was the *threo* (above the plane of the sugar) configuration, were synthesized. Compounding of the *threo*-analogues continued for some time, ceasing only when it became clear that this series produced weak antibacterials and antivirals. Numerous batches of zidovudine were produced for in vivo and toxicity studies. In addition, a number of compounds were synthesized or purchased to extend the structure–activity relationship (SAR) work ongoing at Wellcome. Various analogues with a 3'-substituent different from the azido, such as hydrogen unsaturation (2',3'-didehydro-), and others, like F, Cl, Br, NCS, NH_2, CN, NHC(O)R, were made with thymine, uracil, cytosine, and 5-methylcytosine as the heterocycle. Few analogues retained potent activity. Keeping the sugar moiety as 3'-azido-2',3'-dideoxyribose, changes at the 5-position (H, C_2H_5, C_3H_7, $CH_2CH=BCH_2$, [E-&Z-]$CH=CHBr$, $C=CH$, CF_3, N_3), were investigated. Keeping thymine and the sugar moiety constant, substituents at the 2-, 3-, and 4-positions were varied: 2-(OR, 5, SR, NHR, NR_2); 3-[C(O)Ar, CH_3]; and 4-(OR, OAr, NHR, NR_2, SH, SR).

At the 5'-position, prodrugs to alter the pharmacokinetics of zidovudine were synthesized. The first produced were the mono-, di-, and triphosphates, which serve as metabolic markers since most nucleosides are activated by anabolism. Other 5'-derivatives were the deoxy derivatives and esters (sulfonyl, acyl, arylacyl, and alkoxymethylacyl), homologated and methyl- (and phenyl-)substituted hydroxymethylene and the phosphonate. In the antibacterial mechanistic work, the 5'-esters were labile in the presence of serum in the medium and hence could not be ranked as in the absence of serum. The 5′-deoxy and N-3-methyl derivatives were not active antibacterials or antivirals; this is likely since 5′-phosphate cannot form in the first case, and the N-3-methyl group either interferes with necessary hydrogen bonding in a growing DNA chain or is not a substrate for bacterial or host cell thymidine kinase.

The early HIV tests were limited and difficult to use in the SAR studies at Wellcome. Since each laboratory employed a different assay measuring different parameters (inhibition of HIV-reverse transcriptase, inhibition of p24 viral antigen and syncytia formation), only those samples with distinct activity were tested. Because at that time the development of standardized tests controlling for host cell, infection level, and quantitative measures of activity had not been completed, the activity of a compound was obtained but its potency could not be ranked except relative to zidovudine, which was eventually tested in each system, though not as a standard. Thus, using a general indication of positive or negative activity, the SAR program at Wellcome was extended among the *erythro* and *threo* series using zidovudine as the benchmark compound. Purine analogues were also synthesized and investigated and, while some purine analogues gave interesting results, the risk of cross-resistance deterred further development. When an advantage over zidovudine was discovered, such as longer serum half-life, decreased toxicity, or broader spectrum of antiviral activity, for any one of these compounds, research on that compound would continue. That work is ongoing.

The public revelation of zidovudine's activity in October 1985 placed intense pressure on Wellcome scientists because many groups around the world had precursors to this type of compound and would undoubtedly work to find active analogues. In fact, other investigators worldwide had published results on compounds in the broad category of 2′,3′-dideoxynucleosides, including the 3′-azido-2′,3′-dideoxynucleosides. At Wellcome, analogues were synthesized and extensive study of zidovudine continued with the primary objectives to make, test, and rank the most important compounds.

New synthetic routes to zidovudine were also investigated. An X-ray structure of the compound was obtained, and nuclear magnetic resonance (NMR) studies determined its solution conformation. Its substrate activities with thymidine kinase from bacterial and human host cells were studied. While some analogues were found to be substrates for cellular thymidine kinase, they were not potent against HIV, indicating that the substrate activity for other phosphorylation steps can be rate limiting.

Zidovudine is metabolized to its 5′-O-β-D-glucuronide (GAZT). Although the glucuronide was successfully synthesized in the laboratory, its isolation from the urine of monkeys was less difficult. GAZT is used as an HPLC (high-pressure liquid chromatography) marker in assays of clinical trial samples. The continual need for relatively large amounts of compound in 1985 and 1986 for preclinical and clinical development allowed the chemists to evaluate each of the three synthetic routes referenced in the literature, in order to determine the most productive route, initiate optimization studies, and provide initial recommendations to the development chemists.

The need for large amounts of thymidine, very apparent by March 1985, forced an intense search for producers of thymidine around the world and called for creative measures from all parties concerned with HIV therapies. Compound already produced could be reworked, and the National Institutes of Health (NIH) donated 40 kg of thymidine to allay the crisis. In July 1985 the first pilot plant scale preparation of zidovudine was run, producing usable compound. Since the last two steps of the synthesis required a reaction with azide, which caused some safety concerns at the pilot plant in Greenville, North Carolina, these steps were handled at the Wellcome laboratories at Research Triangle Park in that state. Ultimately (based on some work in 1975 at Wellcome's facilities in Dartford, United Kingdom) safe monitoring for unreacted azide was developed so that the chemical manufacturing department in Greenville has produced the compound since November 1985 without difficulty.

On October 24, 1985 a celebration was held in the chemical development laboratories (CDL) featuring a huge cookie frosted with the zidovudine structure, its melting point, and "Over 15 kg!!!!" The group had completed the final reaction steps on many batches to produce that much clinical-grade material. In January 1986, CDL ran its last development work on zidovudine.

Toxicology Perspective

During its evaluation as a potential antibacterial agent in the early 1980s, the Division of Toxicology and Pathology at Wellcome Research Laboratories was asked by the Department of Microbiology to conduct a preliminary toxicological evaluation of zidovudine. Any "red flags" from a toxicological standpoint that might preclude further work and/or development of the compound as an antibacterial was the main objective at this time. The usual approach at Wellcome is to evaluate the drug in preliminary dose-range-finding (DRF) studies in both rodents and nonrodents. The latter may be purpose-bred beagle dogs or monkeys. With zidovudine, however, there was only enough compound available for a DRF study in rodents. With that in mind, a 2-week oral study in rats in which a variety of antemortem and postmortem parameters were monitored was conducted. The dose levels chosen in this experiment allowed drug conservation but at the same time were significant multiples of the known antibacterial inhibitory concentrations. The results of this study were encouraging in that there were no "red flags" to cause initial concern, but this was not surprising given Wellcome's experience with nucleosides of this class of compounds. So the message given to the chemists and microbiologists based upon completion of this experiment in the early 1980s was that, while there were no significant toxicological findings in this study to preclude the drug's development as an antibacterial, a "final" preliminary assessment would require evaluation in a second species.

By late 1984, the emphasis of the project had changed. The immediate task in support of the development of zidovudine was to conduct the studies considered necessary to support the initial clinical testing of the drug as an antiviral in humans. Beyond that short-term goal, a plan was required for those experiments that would be necessary to support continuing clinical development of the drug and eventually the filing of appropriate regulatory applications. Discussions with clinicians at Wellcome suggested that the initial experience in humans would encompass an intravenous dosing trial in AIDS (acquired immune deficiency syndrome) patients. There were plans to begin oral dosing soon thereafter and, if the drug appeared to be of benefit, patients would be given zidovudine chronically, since it was not likely to be a cure for HIV. With this information in hand the studies thereby needed for the initial Investigational New Drug Application (IND) would include acute (single-dose) intravenous toxicity studies in rats and mice, multiple-dose intravenous toxicity studies in a rodent and a nonrodent species, and an in vitro hemolysis and protein flocculation study in human type O blood. The latter would ensure that the proposed intravenous formulation was compatible with human blood. It was also recognized, however, that oral toxicology studies should be initiated as quickly as possible in order to support that route of dosing clinically.

If these studies were to be conducted in a timely manner, however, there was a major hurdle to be overcome. That hurdle was compound supply. In the Division of Toxicology and Pathology, this required adoption of a much more flexible approach to the design of the supportive toxicology studies. For most drugs, DRFs are conducted to assist in the selection of the dose levels to be employed in formal toxicology studies conducted under Good Laboratory Practices (GLP). The available supply of zidovudine would not support both DRFs and formal toxicology studies. Therefore, the multidose GLP experiments were initiated in the rat and dog without the benefit of preliminary experiments. In the interest of drug conservation, the dose levels for these studies would simply be severalfold multiples of the projected human dose. In consultation with scientists in the Medical and Virology Divisions, the projected human dose was estimated. Based on this projection and the amount of drug available, a 4-week intravenous study in rats at a 10-fold multiple of the projected human dose and a

2-week intravenous study in beagle in which the high dose level was 6-fold greater than the projected human dose could be conducted. Only later was it learned that, unlike the dog, the metabolism of zidovudine in monkeys was identical to that in humans, and for that reason all later nonrodent toxicology studies were carried out in monkeys. In summary, these experiments would provide for an acceptable toxicological evaluation of intravenous zidovudine prior to study of the drug in patients.

These toxicology experiments of zidovudine were mounted and completed in record time. The fact that signed reports for these studies were available 2.5 months after the first animal was dosed was symbolic of the speed with which all scientists and clinicians contributing to the zidovudine effort were approaching their task. Within a few days of the signing of the preclinical toxicology reports, the Phase I clinical trial got under way. This was, however, the last point in the development of zidovudine in which the nonclinical toxicology studies were ahead of the clinical development program.

While the Phase I and Phase II clinical trials proceeded, the Division of Toxicology and Pathology took steps to carry out the remaining studies that were thought to be required for eventual registration. Since the oral route of administration would be the primary route employed in humans, in the remainder of these experiments the drug was given by gavage. One of the primary considerations in determining what studies should be done was the nature of the target population. If zidovudine were to be given to severely ill AIDS patients only, it was felt that very little in terms of additional studies would have been required using an approach for the nonclinical development of the drug which was the same as that employed for the development of potential anticancer drugs. If, however, as suspected, the drug was shown to be efficacious in AIDS patients, clinicians would in all probability eventually target it toward less ill individuals with HIV, and perhaps even HIV-positive asymptomatic persons. So it seemed certain that the risk–benefit equation would likely change over time, and for this reason the decision was made to approach the development of zidovudine as for any compound given to patients chronically. The studies that we considered necessary for a thorough evaluation of the toxicologic potential of zidovudine included the following:

3-month toxicity study in rats	teratology study in rats
3-month toxicity study in rats	perinatal–postnatal study in rats
6-month toxicity study in monkeys	neonatal toxicity study in rats
1-year toxicity study in rats	Ames mutagenicity assay
1-year toxicity study in monkeys	mouse lymphoma cell assay
carcinogenicity study in rats	in vitro cytogenetics assay
carcinogenicity study in mice	in vitro cytogenetics assay
reproduction–fertility study in rats	cell transformation study

Normally, these studies are done in a relatively orderly fashion and in such a way as to support the phase development of the drug under study. Additionally, all of these experiments would usually have been completed prior to submission of the New Drug Application (NDA) for regulatory review. The toxicological development program for zidovudine was a different story altogether. The clinical development of the drug proceeded at such a pace that, whereas the pivotal Phase II trial was halted in September 1986 when it was shown that zidovudine significantly lowered the mortality rate for AIDS patients compared to those receiving placebo, the 6-month toxicology studies in rats and monkeys were still being conducted. The 1-year toxicity studies in rats and monkeys, the oral carcinogenicity studies in rats and mice, and the reproduction–fertility or perinatal–postnatal studies in rats had not yet begun. Those experiments got under way shortly before or near the marketing of the product in March 1987.

Clinical Development of Zidovudine for Treatment of HIV Infection

Wellcome clinicians in the United States first became directly involved in efforts to discover and develop therapy for the treatment of AIDS and associated infections in 1980, a year before the illness was described as a syndrome by the U.S. Public Health Service Centers for Disease Control. Initial involvement was based upon the fact that many of the drugs used to treat the opportunistic infections associated with AIDS are manufactured by Wellcome. In that year, Burroughs Wellcome Co. began to receive a number of calls requesting intravenous Septra® for the treatment of adult patients with *Pneumocystis carinii* pneumonia (PCP). At that time the intravenous preparation of Septra was not marketed in the United States but was available under a treatment IND program. Initially, there was skepticism of the increase in requests because, until that time, episodes of PCP generally occurred primarily in children who had received intensive chemotherapy for leukemia. Approximately a year later, epidemiological studies demonstrated PCP as one of the prime manifestations of AIDS. Wellcome scientists have also studied pyrimethamine (Daraprim®), leucovorin (Wellcovorin®), acyclovir (Zovirax®), DHPG (BW 759U, ganciclovir), and interferon (Wellferon®) to treat various opportunistic infections or tumors occurring in AIDS patients. Because of this involvement, clinicians at Wellcome became familiar with the disease during the early 1980s. In addition Wellcome's long history in the prior two decades of development of antiviral therapy included the development of Marboran® for the prevention and treatment of smallpox and complications of its vaccination in the 1960s, trifluorothymidine (Viroptic®) for ocular herpes infections in the 1970s, and then acyclovir (Zovirax®) for herpes infections in the 1980s.

A Phase I study began in July 1985 and was a collaboration between the National Cancer Institute and Duke University sponsored by Burroughs Wellcome Co. This study was conducted in patients infected with HIV who had been diagnosed as having AIDS or ARC (AIDS-related complex). The results indicated zidovudine was well absorbed orally, with dose-dependent kinetics observed over a fairly wide dosing range. Zidovudine was shown to be 65% bioavailable, but in reality it may be as high as 100% bioavailable. This discrepancy results because there is a first-pass metabolism effect in which a portion of the zidovudine is converted to its 5′-glucuronide as the result of glucuronidation in the liver. Both peak and trough levels that were above the in vitro sensitivity of the virus were achieved in these studies, suggesting that anti-HIV effects might be possible in humans. In addition, it was found that zidovudine penetrated the blood–brain barrier quite well, suggesting that viral infections in brain could be treated. The significant glucuronidation of zidovudine may be an important factor, since other drugs that are glucuronidated may have some effect on its metabolism.

When the Phase I studies were completed in January 1986, a very difficult decision as to how to proceed was required. Traditionally, early clinical studies of new drugs proceed in a very regimented way. New drugs are typically tested in normal, healthy volunteers to evaluate safety and tolerance, and then the new drug is examined in a larger number of patients to evaluate its efficacy and safety profile—usually in patients with milder stages of the disease in question. There are many reasons for this approach. The first is that any toxicity seen is likely to be milder in patients whose baseline physical status is relatively good. More importantly, the likelihood of therapeutic success in less ill patients is often greater in these patients than in those who are at a more severe stage of their disease. In the case of zidovudine, however, Wellcome clinicians believed that there were two counterbalancing elements which required that a less classical approach be taken. The first was that there were a large number of people, possibly hundreds per week, dying of AIDS at the time the Phase I study was completed in January 1986. Wellcome also believed that testing zidovudine in patients with advanced manifestations of HIV infection was the most vigorous test to determine its therapeutic index. If it proved to be effective in the most severely ill patients while exhibiting manageable adverse effects, then it might be more beneficial in patients with milder forms of disease. A difficult decision was therefore made to conduct a double-blind, placebo-controlled study in advanced AIDS and ARC patients in February 1986.

Another key issue in the study was the decision to administer placebo to half of the patients enrolled. This study was initiated at a time when the Phase I study had given only hints that the drug might be effective. Yet with hundreds of people dying and the publication of the Phase I study that described potentially beneficial therapeutic effects in humans, there arose a number of ethical and scientific questions concerning the conduct of a placebo-controlled study. Nevertheless, this drug had to be proven, by classical clinical research methodology, to be both safe and effective, or otherwise many patients might be put at risk without knowledge of the actual benefits of the drug. In order to ensure that the risk and benefits of zidovudine were evaluated adequately without withholding, for any longer than necessary, a promising therapy, it was agreed to appoint a Data Safety and Monitoring Board whose members would examine data from the ongoing study every 2 months and make recommendations about how to proceed. After analysis of various safety and efficacy parameters, this board of medical, ethical, and statistical experts was to advise Wellcome whether one group was experiencing significantly greater side effects or greater benefit from therapy than the other group and recommend whether it would be ethical or not to proceed.

In this study, 282 AIDS and ARC patients were entered at 12 university-associated medical centers in the United States between January and June 1986. In order to have as uniform and comparable groups as possible between patients given drug and those given placebo, narrow categories of disease progression were studied. For the AIDS component, only those patients who had experienced their first episode of PCP within the prior 4 months were entered in the study. ARC patients were enrolled if they had a number of symptoms including, among others, weight loss, sustained fever for over a month, and/or extensive oral candidiasis. All patients were required to have fewer than 500 CD4 cells and to have complete cutaneous anergy to four common antigens. The vast majority of patients had fewer than 200 CD4 cells. Patients with Kaposi's sarcoma, intravenous drug abusers, and children were excluded from this study. The drug and placebo groups were quite comparable in a variety of baseline characteristics that were examined.

On September 19, 1986, the Data Safety and Monitoring Board recommended to Wellcome that the study should be terminated because a significantly higher mortality rate in the placebo group compared to the therapy group was demonstrated. Since patients were enrolled at different times, the length of time on drug ranged from 10 to 28 weeks, with an average of 17 weeks. Analysis of the data at that time indicated that, when compared to placebo, zidovudine recipients had significant improvements in the number of CD4 cells, delayed cutaneous hypersensitivity, weight gain, activities of daily living, and neurological function. In addition, zidovudine recipients had significant decreases (in many cases to an undetectable level) of previously circulating p24 antigen and significant decreases in the frequency and severity of opportunistic infections. Most importantly, the probability of death within 6 months of initiating therapy was 22% for the placebo group and 2% for the drug-treated group.

Symptomatic adverse reactions were extremely common in both groups. This was likely the result of the complicated nature of the underlying disease. Nausea, myalgias, insomnia, and headache, however, were significantly more common in the drug-treated group. The most significant toxicity was myelosuppression, which was dependent upon dose and duration of therapy, as well as upon preexistent bone marrow reserve. Up to 45% of patients with poor bone marrow reserve had significant decreases in either red cell and/or white cell numbers during the observation period. The incidence of such decreases in patients with better but still compromised marrow reserve was only slightly higher than in the same subset of individuals in the placebo group. Such myelosuppression could generally be managed by dose reduction, dose interruption, transfusion, or a combination of these approaches.

At the time the placebo-controlled portion of the study was terminated, all patients, including those originally randomized to receive placebo, were offered the opportunity to receive zidovudine in an unblinded fashion provided they agreed to continued follow-up by the original investigator. While most of the patients agreed to continue taking zidovudine, a small number elected to leave the study for a variety of reasons. Because of these factors, continued follow-up and comparison of the two groups was particularly difficult. Patients can be classified a number of ways depending upon their diagnosis at entry and/or their management after the code was broken. Although, as shown in Table 1, these analyses yield slightly different survival values, there is great consistency in the fact that zidovudine recipients always had a significantly higher survival rate than placebo recipients. Too few patients in the original placebo group remained after 9 months to provide meaningful comparisons. In fact, only 4 of the 28 patients originally assigned to placebo who did not elect to receive zidovudine after unblinding of the study or who received it for less than 3 weeks were alive 1 year after initiation of the study, and all are now dead. Thus, comparison of survival of patients on zidovudine for greater than 9 months must be compared to historical controls. These comparisons are, however, less than ideal because historical groups may represent a significantly different patient cohort and because of the very incomplete follow-up of individual patients in most epidemiological studies used for this comparison. Also, most epidemiological studies use spontaneous reporting of death or registration of death certificates specifying death from AIDS within a particular locale to make their projection of survival rates. Both of these factors lead to significant underreporting of deaths.

TABLE 1. Extended survival of patients in original double-blind, placebo-controlled study

Patient category	Months				
	6	9	12	15	18
1. All original Retrovir recipients	96.5	91.5	84.3	78.6	68.3
2. Same as 1, but excluding patients off drug >60 days before death	97.7	93.7	88.7	85.2	77.7
3. Same as 1, but prophylaxed at any time for PCP	100.0	97.7	90.7	90.7	83.5
4. Same as 1, but never prophylaxed for PCP	94.9	88.8	81.5	74.2	61.5
5. Same as 1, but AIDS only	95.3	90.2	78.8	74.0	61.8
6. Same as 5, but calculated since date of PCP rather than date of entry into study	100.0	92.9	87.1	83.5	72.8
7. Same as 1, but ARC only	98.2	96.4	92.8	85.5	78.2
8. Placebo recipients who received no AZT or <3 weeks AZT	75.8	51.5	–	–	4.1[a]
9. Same as 8, but AIDS only	69.2	38.8	22.2	–	0–4.1[a]
10. Same as 8, but ARC only	83.2	67.8	–	–	0–4.1[a]

[a] Only one patient alive at 18 months in "true placebo" group. Unclear at this time whether AIDS or ARC. This patient died 2 months later. Follow-up between 9 and 18 months unclear because patients dropped out of study.

For example, a study that made extensive efforts to track down purported "longer term survivors" of AIDS found that at least 58% of such patients were in fact, dead.

With these caveats in mind, the best historical comparison to the original cohort of patients who were randomized to zidovudine in the study is a cohort of AIDS patients in New York City in 1985 who had their diagnosis made exclusively on the basis of PCP. Their minimum 1-year mortality from the date of PCP diagnosis was 51%. Mortality was probably significantly greater because only those patients who had AIDS listed as their cause of death on a death certificate registered in New York City were considered to be dead for purposes of the study. Because AIDS patients in the original double-blind, placebo-controlled study began taking zidovudine about $2\frac{1}{2}$ months after their original episode of PCP, their 13% mortality at 1 year after diagnosis of PCP and 21% after entry into the study is one-fourth to one-half the minimal mortality reported in the New York study.

The denouement of the original double-blind, placebo-controlled study in September 1986, provided another opportunity to study a large cohort of patients. Wellcome set up a program, in conjunction with the National Institutes of Health, to dispense zidovudine free of charge to any AIDS patient in the United States who had had PCP at any time in the past and who fulfilled minimum entry criteria. ARC patients were not included in this program. Approximately 4800 AIDS patients received AZT under this "treatment IND," "compassionate plea," or "parallel track" program between October 1986 and March 1987, when the drug became available by prescription. The characteristics of the patients in this study were similar to those of AIDS patients in the general population, with the vast majority being homosexual or bisexual men. Nevertheless, a number of patient categories not well represented in the Phase II double-blind, placebo-controlled study did participate in this uncontrolled study. There were nearly 150 women and more than 250 intravenous drug abusers. In addition, 424 patients were Hispanic and more than 500 were black. Although zidovudine was approved for general use in March 1987, the survival of these patients could be monitored until September 15, 1987, when the controlled distribution program that was in place during that interval was dismantled. The amount of data that can be obtained during "treatment IND" studies is generally limited, but sufficient controls were instituted so that mortality statistics are reasonably reliable, at least for a 9-month period. After adjusting for the fact that significantly sicker patients could participate, overall survival data were very similar to those observed in the original placebo-controlled study.

The incidence of adverse reactions was somewhat less than noted in the original placebo-controlled study and may have been the result of less intensive observation and management of the patients or less aggressive reporting of such reactions. Although significantly higher rates of death have been reported in untreated women and drug addicts with AIDS when compared to male homosexuals with AIDS, no such differences were noted if these patients were receiving zidovudine. Likewise, no differences in mortality were noted between black and white AIDS patients receiving zidovudine. The highest mortality rates were recorded during the first 8 weeks of study, indicating the very advanced state of illness of many of the participants. The mortality of people who had acquired disease through blood transfusion was somewhat higher than those who had acquired infection by other means. This observation may have been the result of the more advanced age of such patients, as well as their poorer general state of health.

Certain prognostic factors of survival were noted in this study (Table 2). Better survival was associated with higher hemoglobin and performance levels (Karnofsky Score) at enrollment, as well as the brevity of the period between the first episode of PCP and the initiation of zidovudine therapy. These data point to the importance of beginning therapy as soon as possible after the diagnosis of AIDS or advanced ARC is made.

Although a great deal of information about the usefulness of zidovudine was gathered in a relatively short period of time, a very aggressive worldwide program of clinical research was mounted to address many

TABLE 2. Treatment IND study

Factors not associated with difference in survival while on AZT:	Factors associated with difference in survival while on AZT:
Sex	Baseline hemoglobin
Race	Baseline performance level of activities of daily living (Karnofsky Score)
Method of virus acquisition[a]	Duration since diagnosis of PCP

[a]With the exception of blood transfusion recipients.

unanswered questions. In the United States this program was conducted, in part, in conjunction with the AIDS Treatment and Evaluation Units (ATEUs), now the AIDS Clinical Trials Group (ACTG), of the National Institutes of Health. The largest group of studies involved patients with different degrees of severity of HIV infection, including patients with advanced disease (AIDS), milder forms of ARC, lymphadenopathy syndrome and those who are infected but who do not have obvious signs or symptoms of disease. The results of these landmark clinical trials have defined the role of zidovudine in treating all stages of HIV infection and disease. Consequently, zidovudine has emerged as the cornerstone of therapy for HIV infection and the most commonly prescribed initial treatment, as well as the agent with which new therapies are usually compared.

Studies have been conducted in special patient populations such as hemophiliacs, intravenous drug users, and children. Data from children indicate that their absorption, distribution, metabolism, and excretion of zidovudine is similar to those of adults, as are the benefits and adverse reactions to the drug. Particularly striking improvement in neurological function have been noted in pediatric patients. Additional studies concerning the effect of zidovudine on abnormal neurological function in HIV-infected adults have been conducted. Although significant improvements in neurological function in adults with AIDS and ARC have already been noted in controlled as well as uncontrolled studies, additional research is required to define precisely the degree of benefit and to understand the occasional neurological adverse experiences associated with the use of zidovudine. Likewise, the role of zidovudine in improving the thrombocytopenia often seen in AIDS patients has been evaluated.

The use of zidovudine in conjunction with a variety of other medications is being examined for two distinct reasons. Certain drugs such as other nucleoside analogues like ddC (2′,3′-dideoxycitidine), ddI (2′,3′-dideoxyinosine), acyclovir, and interferon have been shown to be synergistic in vitro with zidovudine in inhibiting HIV replication (Hartshorn et al., 1987; Mitchell et al., 1987; Ruprecht et al., 1987), but additional studies are required to determine whether an additive or synergistic effect can be observed in people. Additionally, some compounds such as G-CSF and erythropoietin counteract the marrow suppressive effects of zidovudine. Some immunomodulators may enhance immune function at the same time that zidovudine inhibits viral replication, and such combinations are currently under study.

Studies have been conducted to determine the safety and tolerance of zidovudine when used in conjunction with drugs employed in the therapy of opportunistic infections. In addition, studies of the appropriate dosage adjustment in patients with renal and/or hepatic failure have been included as part of an extensive postmarketing surveillance program. Finally, intensive viral sensitivity studies are continuing to determine the significance of resistance development. Although years may pass before the results of some of these studies enable us to have a more complete knowledge of the full therapeutic profile of zidovudine, sufficient data already exist to indicate that the drug is a valuable weapon in the physician's armamentarium to improve and lengthen the life of patients with HIV infection.

Acknowledgments

Contributions were made to this chapter by Ms. Kathy Bartlett, Drs. Kenneth M. Ayers, Gertrude B. Elion, Phillip A. Furman, Janet L. Rideout, and Marty St. Clair of the Wellcome Research Laboratories.

References

Ayers KM (1987): Preclinical toxicology of zidovudine. *Wellcome Int. Antiviral Symp, Monte Carlo, 1987*

Bach MC (1987a): Zidovudine for lymphocytic interstitial pneumonia associated with AIDS. *Lancet* 2:796

Bach MC (1987b): Possible drug interaction during therapy with azidothymidine and acyclovir for AIDS. *N Engl J Med* 316:547

Barre-Sinoussi F, Chermann JC, Rey F, et al. (1983): Isolation of a T-lymphotropic retrovirus from a patient at risk for acquired immune deficiency syndrome (AIDS). *Science* 220:868–71

Barry DW (1986): Testimony before the Intergovernmental Relations & Human Resources Subcommittee of the Committee on Government Operations, The Honorable Ted Weiss, Chairman. U.S. House of Representatives, Washington, DC (July 1).

Blanche S, Rouzioux C, Caniglia M, Tardieu M, Griscelli C (1987): Zidovudine in eight HIV-infected children for a 6-month period. *Wellcome Int Antiviral Symp, Monte Carlo, 1987*

CDC (1981a): Kaposi's sarcoma and *Pneumocystis* pneumonia among homosexual men—New York City and California. *Morbid Mortal Wkly Rep* 30:305–308.

CDC (1981b): Follow-up on Kaposi's sarcoma and *Pneumocystis* pneumonia. *Morbid Mortal Wkly Rep* 30:409–410

CDC (1987): Diagnosis and management of mycobacterial infection and disease in persons with Human Immunodeficiency Virus infection. *Ann Intern Med* 106:254–256

Chaisson RE, Allain JP, Leuther M, Volberding PA (1986): Significant changes in HIV antigen level in the serum of patients treated with azidothymidine. *N Engl J Med* 315:1610–1611

Davtyan DG, Vinters HV (1987): Wernicke's encephalopathy in AIDS patient treated with zidovudine. *Lancet* 1:919–920

Devita VT, Broder S, Fauci AS, Kovacs JA, Chabner BA (1987): Developmental therapeutics and the acquired immunodeficiency syndrome. *Ann Intern Med* 106:568–581

Eggleston M (1987): Clinical review of ribavirin. *Infect Control* 8:215–218

Elwell LP, Ferone R, Freeman GA, et al. (1987): Antibacterial activity and mechanism of action of 3'-azido-3'-deoxythymidine (BW A509U). *Antimicrob Agents Chemother* 31:274–280

Fischl MA, Richman DD, Grieco MH, et al. (1987): The efficacy of azidothymidine (AZT) in the treatment of patients with AIDS and AIDS-related complex. A double-blind, placebo-controlled trial. *N Engl J Med* 317:185–191

Forester G (1987): Profound cytopenia secondary to azidothymidine. *N Engl J Med* 317:772.

Furman PA, Fyfe JA, St Clair MH, et al. (1986): Phosphorylation of 3'-azido-3'-deoxythymidine and selective interaction of the 5'-triphosphate with human immunodeficiency virus reverse transcriptase. *Proc Natl Acad Sci USA* 83:8333–8337

Gill PS, Rarick M, Brynes RK, Causey D, Loureiro C, Levine AM (1987): Azidothymidine associated with bone marrow failure in the Acquired Immunodeficiency Syndrome (AIDS). *Ann Intern Med* 107:502–505

Glinski RP, Khan MS, Kalmas RL, Sporn MB (1973): Nucleoside synthesis. IV. Phosphorylated 3'-amino-3'-deoxythymidine and 5'-amino-5' deoxythymidine and derivatives. *J Org Chem* 38:4299–4305

Gottlieb MS, Wolfe PR, Chafey S (1987): Case report: Response of AIDS-related thrombocytopenia to intravenous and oral azidothymidine (3'-azido-3'-deoxythymidine).*AIDS Res Hum Retroviruses* 3:109–114

Hammer SM, Gillis JM (1987): Synergistic activity of granulocyte-macrophage colony-stimulating factor and 3'-azido-3' deoxythymidine against human immunodeficiency virus *in vitro. Antimicrob Agents Chemother* 31:1046–1050

Hartshorn KL, Vogt MW, Chou TC, et al. (1987): Synergistic inhibition of human immunodeficiency virus *in vitro* by azidothymidine and recombinant alpha A interferon. *Antimicrob Agents Chemother* 31:168–172

Horsburgh CR, Mason UG, Farhi DC, Iseman MD (1985): Disseminated infection with *Mycobacterium avium intracellular:* A report of 13 cases and a review of the literature. *Medicine (Baltimore)* 64:36–47

Horwitz JP, Chua J, Noel M (1964): Nucleosides. V. The monomesylates of 1-(2'-deoxy-β-d-lyxofuranosyl) thymine. *J Org Chem* 29:2076–2078

ICN (1986): *Pharmaceuticals: Virazole Product Monograph.* Costa Mesa, CA: ICN.

Jollow PJ, Thorgeirsson SS, Potter WZ, Hashimoto M, Mitchell JR (1974): Acetaminophen-induced hepatic necrosis. VI. Metabolic disposition of toxic and nontoxic doses of acetaminophen. *Pharmacology* 12:251–271

Klecker RW, Collins JM, Yarchoan R, et al. Plasma and cerebrospinal fluid pharmacokinetics of 3′-azido-3′-deoxythymidine: A novel pyrimidine analog with potential application for the treatment of patients with AIDS and related diseases. *Clin Pharmacol Ther* 41:407–412

Krenitsky TA, Koszalka GW, Tuttle JV, Rideout JL, Elion GB (1981): An enzymatic synthesis of purine D-arabinosides. *Carbohydr Res* 97:139–146

Lin TS, Prusoff WH (1978): Synthesis and biological activity of several amino analogues of thymidine. *J Med Chem* 21:109–112

Lyerly HK, Cohen OJ, Weinhold KJ (1987): Transmission of HIV by antigen-presenting cells during T-cell activation: Prevention by 3′-azido-3′-deoxythymidine. *AIDS Res Hum Retroviruses* 3:87–94

Mitchell WM, Montefiori DC, Robinson WE, Strayer DR, Carter WA (1987): Mismatched double-stranded RNA (Ampligen) reduces concentration of zidovudine (azidothymidine) required for *in vitro* inhibition of human immunodeficiency virus. *Lancet* 1:890–892

Mitsuya H, Broder S (1987): Strategies for antiviral therapy in AIDS. *Nature (London)* 325:773–778

Mitsuya H, Weinhold KS, Furman PA, et al. (1985): 3′-Azido-3′-deoxythymidine (BW A509U): An antiviral agent that inhibits the infectivity and cytopathic effect of human T-lymphotropic virus type III/lymphadenopathy-associated virus *in vitro*. *Proc Natl Acad Sci USA* 82:7096–7100

Mitsuya H, Matsukura M, Broder S (1987): Rapid *in vitro* systems for assessing activity of agents against HTLV-III/LAV. In: *AIDS: Modern Concepts and Therapeutic Challenges*, Broder S, ed. New York: Dekker, pp 303–333

Nakashima H, Matsui T, Harada S, et al. (1986): Inhibition of replication and cytopathic effect of human T-cell lymphotropic virus type III/lymphadenopathy-associated virus by 3′-azido-3′-deoxythymidine *in vitro*. *Antimicrob Agents Chemother* 30:933–937

Pizzo PA (1987): A Phase I study of zidovudine administered as a continuous intravenous infusion to children with AIDS and AIDS-related complex. *Wellcome Int Antiviral Symp, Monte Carlo, 1987*

Popovic M, Sarngadharan MG, Read E, Gallo RC (1984): Detection, isolation, and production of cytopathic retroviruses (HTLV-III) from patients with AIDS and pre-AIDS. *Science* 224:497–500

Richman DD, Fischl MA, Grieco MH, et al. (1987): The toxicity of azidothymidine (AZT) in the treatment of patients with AIDS and AIDS-related complex. A double-blind, placebo-controlled trial. *N Engl J Med* 317:192–197

Rothenberg R, Woelfel M, Stoneburner R, Milberg J, Parker R, Truman B (1987): Survival with the acquired immunodeficiency syndrome: Experience with 5833 cases in New York City. *N Engl J Med* 317:1297–1302

Ruprecht RM, O'Brien LG, Rossoni LD, Nusinoff-Lehrman S (1986): Suppression of mouse viraemia and retroviral disease by 3′-azido-3′-deoxythymidine. *Nature (London)* 323:467–469

Ruprecht T, O'Brien L, Rosas D, Andersen J (1987): Recombinant human interferon alpha A/D (RHulfn-alpha A/D) enhances the antiretroviral effect of 3′-azido-3′-deoxythymidine (AZT) in mice. *Proc Am Assoc Cancer Res* 28:456

Sharpe AH, Jaenisch R, Ruprecht RM (1987): Retroviruses and mouse embryos: A rapid model for neurovirulence and transplacental antiviral therapy. *Science* 236:1671–1674

St Clair MH, Richards CA, Spector T, et al. (1987): 3'-Azido-3'-deoxythymidine triphosphate as an inhibitor and substrate of purified human immunodeficiency virus reverse transcriptase. *Antimicrob Agents Chemother* 31:1972–1977

Tavares L, Roneker C, Johnston K, Nusinoff-Lehrman S, DeNoronha F (1987): 3'-Azido-3'-deoxythymidine in feline leukemia virus–infected cats: A model for therapy and prophylaxis of AIDS. *Cancer Res* 47:3190–3194

Vogt MW, Hartshorn KL, Furman PA, et al. (1987): Ribavirin antagonizes the effect of azidothymidine on HIV replication. *Science* 235:1376–1379

Yarchoan R, Klecker RW, Weinhold KJ, et al. (1986): Administration of 3'-azido-3'-deoxythymidine, an inhibitor of HTLV-III/LAV replication to patients with AIDS or AIDS-related complex. *Lancet* 1:575–580

Yarchoan R, Berg G, Brouwers P, et al. (1986): Response of human immunodeficiency virus-associated neurological disease to 3'-azido-3'-deoxythymidine. *Lancet* 1:132–135

3

Discovery of Nevirapine, a Nonnucleoside Inhibitor of HIV-1 Reverse Transcriptase

JULIAN ADAMS AND VINCENT J. MERLUZZI

Introduction

Five years after the description of acquired immunodeficiency syndrome (AIDS) and approximately 3 years after the discovery of human immunodeficiency virus type 1 (HIV-1), we began a study group around the therapeutic target of HIV-1 protease (a virus-specific enzyme involved in processing and maturation). The goals of this group were to study the HIV-1 protease, its substrates, and use of possible peptide inhibitors. During this time a few of us discussed the possibility of targeting HIV-1 reverse transcriptase (RT) as a target for therapeutic intervention. It was clear that resources could not be spread too thin and it was also clear to us that we did not want to design or rediscover nucleoside analogues as inhibitors. In fact, if we were to embark on an inhibitor program for RT, our foremost objective was to search for and design inhibitors that were clearly nonnucleoside in nature. RT seemed to be an excellent target because it was known to be unique to retroviruses (an exception being hepatitis B) and not usually present in normal cells. Yet, the only true worthwhile inhibitors of this enzyme were nucleoside analogues of which AZT (zidovudine) was the prototype and soon after the only approved drug for HIV-1. It was clear to some of us that we should somehow target HIV-1 RT but in a way that resources, equipment, and personnel use did not detract from ongoing projects and in particular the HIV-1 protease research team. Since this was a chance venture, it was decided that we would design an RT "screen" using an established assay system and randomly test all compounds available in the worldwide Boehringer Ingelheim compound repositories. If any true positive compounds were identified as inhibitors, we would then rationally

The Search for Antiviral Drugs
Julian Adams and Vincent J. Merluzzi, Editors
© 1993 Birkhäuser Boston

search for analogues and start structure–activity relationship (SAR) based synthesis here in the United States. It was necessary from the beginning to identify which RT to use and how to reassure ourselves of finding a true inhibitor, that is, specific for RT only. We decided from the very beginning not to address virus replication but rather to concentrate on inhibitors of the RT enzyme and address the question of viral replication later. This meant, of course, that an inhibitor, once discovered, might or might not work on the virus. Problems that could arise might be an inability to penetrate the cell, solubility in culture medium, cell metabolism of the compound, and generally unforseen metabolic problems. Yet, we felt that if a specific RT inhibitor could be found, cell penetration and other problems could be overcome by synthetic design.

The Screen

A proposal was written and accepted by management to begin the random screening process for inhibitors of RT. It was decided that two people would begin the process, a laboratory supervisor and a laboratory scientist. Several mistakes were made in the beginning. The most time-consuming and costly was the use of the murine RT enzyme to begin the screening process. We had realized that the HIV-1 RT was the best and most appropriate enzyme, but it was not readily available to us at that time. The murine enzyme did, however, allow us to begin the process of screening large numbers of compounds and automate the process. We had decided that one person could process approximately 200 random compounds per week at one dose. In addition, follow-up enzyme specificity tests (in the beginning, *Escherichia coli* DNA polymerase), data management, and reagent stocks could be handled by the same person. After several hundred tests, a compound was found that inhibited the murine RT enzyme at 30 μM while having no effect on *E. coli* DNA polymerase. The compound was a nonnucleoside, and several analogues were also active but to a lesser extent. It was imperative at this point to test the activity of this inhibitor on HIV-1 RT. We obtained particle-derived HIV-1 RT, but unfortunately this compound was completely inactive on the HIV-1 RT enzyme. Also, there was no activity against HIV-1 virus replication as measured by syncytia formation by our collaborator, Dr. John Sullivan at the University of Massachusetts Medical School. Although the compound was inactive, it was at this point that the research group expanded from two biologists to five researchers, including a medicinal chemist, an enzymologist, and a murine retrovirologist.

After some thought it was decided that the only approach to take was to obtain the clone for HIV-1 RT (several were eventually obtained), express the protein, purify it, and use this material for screening. The murine screen was eventually dropped and replaced with the HIV-1 RT screen after some early kinetics, pH optimization, ionic strength, and substrate-saturating kinetics were performed. It was clear that while the murine RT enzyme helped us set up the mechanics (e.g. automation) of a high-capacity screen, the murine RT enzyme was significantly different from the HIV-1 RT enzyme.

Simultaneous to the development of the HIV-1 RT screen, we decided to have on hand other retrovirus RT enzymes to test for specificity; these included the avian, simian, and feline (FeLV) RT enzymes. Another specificity screen would include the calf thymus DNA polymerase and eventually the human DNA polymerases (alpha, beta, delta, and gamma). The latter enzymes and expertise were kindly provided by Dr. Y.-C. Cheng at Yale University. If any positive compounds were found with sufficient potency, our collaborator Dr. John Sullivan would provide immediate tests of the compound for HIV-1 replication using both syncytia and p24 measurements.

The First Leads

In September 1988 we began screening 200 random compounds per week against HIV-1 RT. Our hopes were to obtain an approximate 0.1% "hit" rate and that a certain percentage of those hits would initially pass an enzyme specificity battery of tests, most importantly the mammalian DNA polymerases. At this point, the working group consisted of five researchers but mainly a scientist performing the screening assays and data management. Three months after the screening began a compound was found that inhibited the HIV-1 RT enzyme with an IC_{50} of 6 μM. This first active compound was related to a group of anti-muscarinic receptor analogues synthesized many years ago and related to pirenzepine (Gastrozepin®), an antiulcer compound marketed by Boehringer Ingelheim in Europe (Figure 1). In follow-up tests this compound had no effect on other RT enzymes and had no effect on calf thymus DNA polymerase. The potency was weak but gave us a "departure" point for analogue searching within the Boehringer Ingelheim compound repository. We opted not to test this compound against viral replication but rather to obtain as many structural analogues as possible to test for an increase in potency. The process of random screening would continue along with analogues of the first active. (Implicit in the remaining discussion is that the random screening process never halted, but for the sake of simplicity we will focus on the follow-up to the first active, which in fact led to the discovery of our development compound.) As a

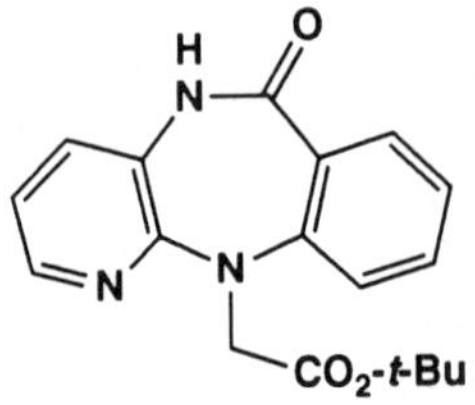

IC_{50} = 6 μM

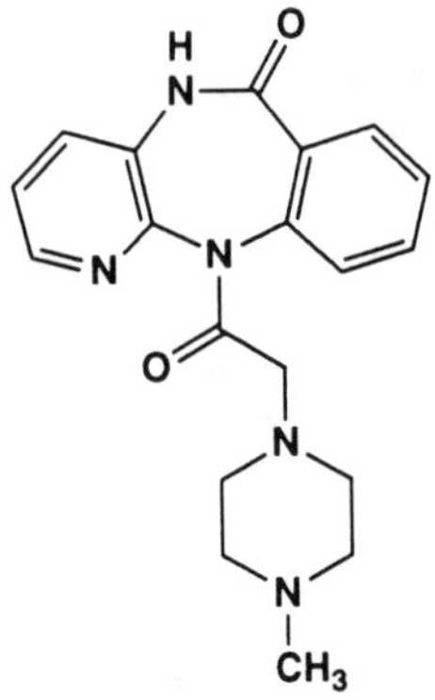

Pirenzepine
IC_{50} >> 10 μM

FIGURE 1. *First lead:* Tricyclic series was discovered, related to the anti-muscarinic pirenzepine.

result of this previous program, we had over a thousand analogues to test, but within 3 months an analogue with an IC_{50} of 350 nM was found in the series. This compound (dubbed "LS") was inactive against all specificity enzymes and active in cell culture against HIV-1 (Figure 2). With only scant information on LS and because of general skepticism, many colleagues and some of management had reservations about the true antiviral nature of this compound. Several experiments that showed no toxicity to uninfected cells and no direct effect against syncytia [cytopathic effects (CPE)] by other viruses helped but did not sway the purists. Since the potency was reasonable, several comments were made about developing LS as far as one could while still searching for better analogues and beginning a medicinal chemistry SAR program around LS. At this same time, a research group and a development group were formed and included areas outside of preclinical. A final

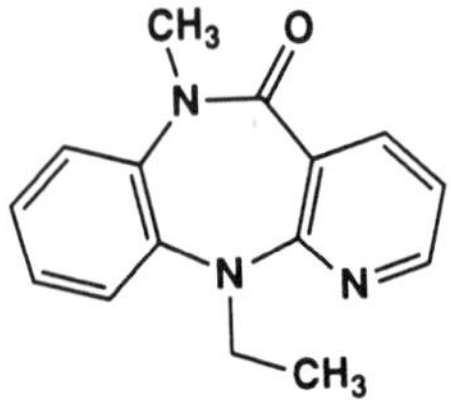

FIGURE 2. LS compound (IC_{50} = 350 nM).

decision was made not to develop LS but to obtain as much data as possible on it, including solubility, metabolism, virus specificity, toxicity, pharmacology, and preliminary formulation/stability data. The search for a development candidate with acceptable potency, metabolism, solubility, pharmacokinetics, synthesis, and pharmacology would be included in the SAR. All of these requirements were simultaneously part of the screening process to choose a development candidate based on the LS lead. We felt that while this was a large undertaking, it would be the only way to move quickly and arrive at an optimal development candidate. It was clear to many of us that moving prematurely with an inferior compound would set the development process back.

Choosing a Development Candidate

Strategic Considerations

The LS compound had an IC_{50} value of 350 nM in our standard RT inhibition assay and also demonstrated good antiviral activity (inhibition of p24 and syncytial formation in C8166 T lymphocytes). Early consideration to develop this compound as a clinical candidate prompted a discussion as to what selection criteria should be adopted. In the early days of the program, the chemists argued that the potency could be improved. Though a thorough substructure screening of tricyclic compounds in the Boehringer sample collection had been undertaken, synthesis of new compounds had only just begun. The synthetic efforts were expanded to quickly identify important structure–activity relationships (SAR) and develop general synthetic methodology to prepare targeted analogues related to the LS compound. Target values of <100nM IC_{50} in both the RT enzyme and HIV viral replication assays were arbitrarily chosen as suitable potency benchmarks.

Specificity of all potent new analogues was monitored with respect to other reverse transcriptases and mammalian DNA polymerases. In this regard we were gratified at the exquisite specificity of the tricyclic compounds, since none of the related polymerase enzymes were inhibited. This inspired confidence that our inhibitors were recognizing HIV-1 RT in a unique and specific manner. Of greater concern was the potential interaction of our inhibitors with other enzymes and receptors, which might eventually cause undesired side effects during antiviral therapy. To this end, all potent compounds were put through a general pharmacological evaluation and were sent to Biomeasure Inc. (Boston) for screening in a panel of receptor-binding assays. We were particularly concerned with possible residual muscarinic receptor binding and possible activity at the benzodiazepine receptors. Happily we found that the class of tricyclic diazepinones of interest proved to be devoid of any ancillary pharmacology, as determined by our screening protocols.

A unique feature of this process of developing a drug for HIV infection was that animal models were not available to test the efficacy of our analogues. Thus particular care was given to the pharmaceutical properties and biodistribution of our drug candidates. We decided to systematically evaluate the solubility, oral absorption, and metabolic stability of our most potent analogues with these criteria in mind. It was determined that a suitable solubility of 100 μg/mL in water at pH 7 was a good target. Our compounds were all weak bases and so, we assumed, would be well dissolved in the acidic environment of the stomach following oral administration. Without the benefit of an efficacy readout we measured the oral absorption of drugs by measuring the concentration in plasma and relating it to the antiviral IC_{50} value determined in cell culture assays. Thus the goal was to maximize the IC_{50} equivalents observed in the plasma as a function of both peak levels and duration. We set as an arbitrary value that a minimum of 10 IC_{50} equivalents at 8 hours would be predicted to have sufficient antiviral protection. We also examined the concentration of drug in other relevant tissues such as the brain, since it is well documented that HIV has significant involvement in the central nervous system. As expected, our weakly basic tricyclic diazepinones readily penetrated the blood–brain barrier. Finally we sought to identify the metabolic fate of our drugs and use this information to block undesired conversion to inactive metabolites. In many cases the structures of metabolites were identified and chemically synthesized and further evaluated in antiviral and pharmacological screens. Ultimately, we worked toward a compound that would be devoid of metabolic inactivation and a parent drug that would persist with a reasonable half-life of at least 8 hours.

Finally the issue of safety and toxicology presented itself as the ultimate hurdle in the development of a new therapeutic. Evaluation of drug analogues in the Ames test for mutagenicity was performed to ensure the general safety of the tricyclic compounds. Some measure of safety was determined by single-dose oral administration in our pharmacokinetic studies. Since safety assessment is time consuming, we opted to evaluate several potent analogues by performing "mini-safety" experiments. This involved a 2-week daily dosing protocol in rats and dogs. This enabled us to compare safety profiles with several compounds as well as to gain experience to anticipate future dosing regimens.

The above guidelines helped guide both the discovery and development phases of the program. Many of the activities that are normally performed sequentially in a drug discovery program were performed in parallel in our search for an RT inhibitor. It was clear from the outset that time was limited, and the research team certainly felt the pressure to streamline activities and even proceed at times with incomplete data to expedite the discovery process. This required an enormous level of cooperation as well as sufficient resources to perform the necessary chemical synthesis, testing, and in vivo evaluation in a relevant time frame so that a meaningful decision could be adopted. Everyone realized what was at stake and supported the efforts in the program.

Chemistry

The chemical synthesis program was driven largely as a result of empirical observations that included the screening of the diazepinones from the sample collection and a systematic probing for pharmacophore substitution patterns on the tricyclic skeleton. The biochemists and the chemists worked closely to test inhibitors in the enzyme assay with a rapid turnaround time for data so that critical decisions guiding synthesis could be implemented. The enzyme assay was a straightforward polymerase assay employing a poly rC RNA template and oligo dG DNA primer and [^{3}H]dGTP as substrate. The assay protocol is shown in Figure 3.

Both the first lead compounds and the LS compound were monopyridodiazepinone derivatives. A systematic comparison of nitrogen atom placement in the A and C rings was undertaken (Figure 4), and an obvious pattern emerged indicating that as a general rule the dipyridodiazepinones were found to be the most potent compounds. This finding was matched by the observation that the dipyrido series was also more soluble, owing to the additional nitrogen atom, and less prone to metabolic *N*-dealkylation (*vide infra*). We next looked for positional specificity of nitrogen atoms in the A and C rings and found that only the 1 and 10 positions afforded good inhibitory activity and even corresponding pyrimidine derivatives were devoid of RT inhibitory activity. (Figure 5).

As a general finding optimal substituents at the amide nitrogen and the N-11 central ring nitrogen were small aliphatic alkyl chains, with the best compound being BI-RH-397 (IC_{50} = 125 nM). This modest improvement over the LS compound was accompanied by another serious problem related to the metabolic stability of our lead compound (Figure 6). *N*-Dealkylation, traced to enzymatic *P*-450 oxidation, led to rapid *N*-demethylation and *N*-deethylation in rats as well as in human liver microsomal preparations. Demethylation per se was not a serious concern since the free amide was itself a good inhibitor (IC_{50} = 475 nM), but deethylation at N-11 led to an inactive compound. We focused our attention on finding a better substituent at the central ring nitrogen, and the cyclopropyl group was introduced and found to be a relatively stable substituent with compound BI-RG-580 showing similar potency (IC_{50} =

HIV-1 RT	**p66/p51 heterodimer** **- from Summer clone (Yale University)** **- purity >95%**
Template:Primer	**poly rC : oligo dG**
dNTP	^{3}**H-dGTP (6 μM)**
Reaction mixture **(final concentrations)**	**0.7 nM enzyme** **50 mM Tris** **42 mM KCl** **4.8 mM DTT** **3.6 mM MgCl$_2$** **0.25 mg/mL HSA** **pH 7.8, room temperature**
Procedure	**Add inhibitor to reaction mixture** **Wait 5 min.** **Add substrate** **Wait 60 min.** **Add CCl$_3$CO$_2$H to ppt. polynucleotides** **Wait 15 min. at 4 °C** **Filter** **Measure counts**

FIGURE 3. HIV-1 reverse transcriptase screening assay: DTT, dithiothreitol; HSA, human serum albumin.

120 nM) to the *N*-ethyl analogue BI-RH-397. The reason for this improved resistance to *P*-450 oxidation is not clear, but we speculated that the partial sp^2 character of the cyclopropane may lower the oxidation potential of the adjacent nitrogen.

Next we examined the substitution requirements in the A and C rings. A systematic exercise, probing for activity, of placing a methyl group at each position of the pyridine rings was revealing. The results are summarized in Figure 7. Positions 3, 7, 8, and 9 showed a deleterious effect on the ability of analogues to inhibit RT. Position 2 offered neither an enhancement nor a lowering of activity. The most interesting findings resulted from substitution at position 4. With the amide nitrogen bearing a methyl group (the optimal substituent at that position) a 4-methyl substitution proved to be unsuitable for good inhibitory activity (BI-RH-

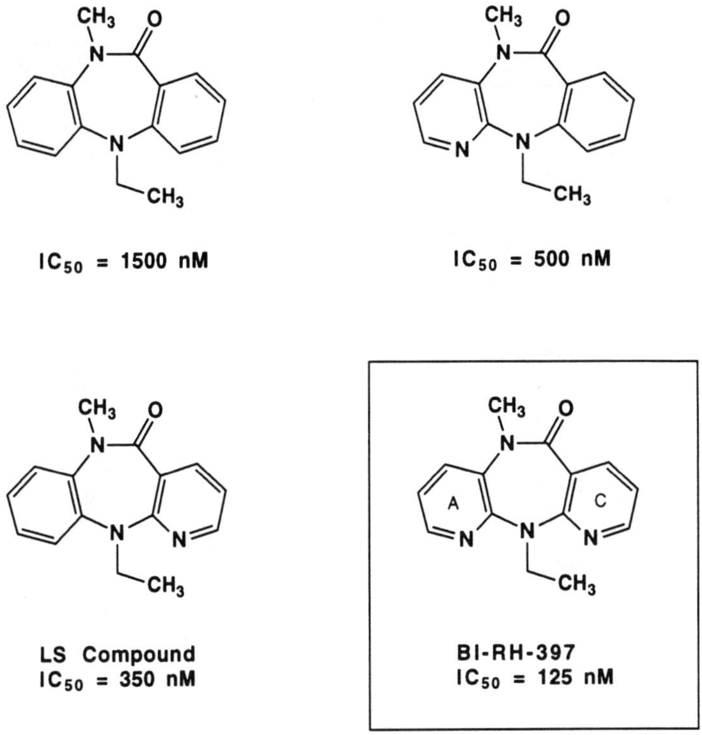

IC_{50} = 1500 nM

IC_{50} = 500 nM

LS Compound
IC_{50} = 350 nM

BI-RH-397
IC_{50} = 125 nM

FIGURE 4. The dipyridodiazepinones are both more potent and more soluble than the corresponding monopyrido- or dibenzodiazepinones.

411, IC_{50} = 865 nM). However, if the amide was unsubstituted (R^1 = hydrogen), a remarkable enhancement of inhibitor activity was observed (BI-RH-414, IC_{50} = 35nM) (Figure 8). The same observation was made with N-11 cyclopropyl derivatives BI-RG-588 and BI-RG-587 (nevirapine), respectively.

This remarkable increase in potency of our drug was attributed solely to the empirical observation that the enzyme recognizes a small alkyl (methyl is optimal) in the "northern" part of the molecule for good binding affinity. We rationalized that either N-methyl or C-4-methyl offers a good fit for RT but the presence of two methyl groups (i.e., BI-RH-411) interferes with optimal binding to the enzyme. Molecular models of the N-11 cyclopropyl derivatives and single crystal X-ray analysis indicated a steric buttressing of the two methyl groups causing a distortion of the central ring amide in the tricyclic dipyridodiazepinones. This leads to poorer binding geometry on the enzyme.

Further modification of substituent placement in the pyridine rings incorporating electron-withdrawing or electron-donating groups to alter the electronic character of the analogues did not offer any improvements over the optimal 4-methyl substituent.

Dipyridodiazepinones

Pyrimidopyridodiazepinones

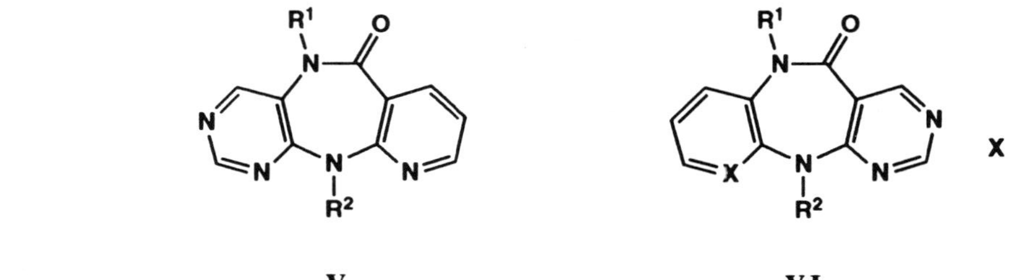

FIGURE 5. Position of nitrogen in the aromatic rings. Dipyridodiazepinones **IV** are most active against HIV-1 reverse transcriptase.

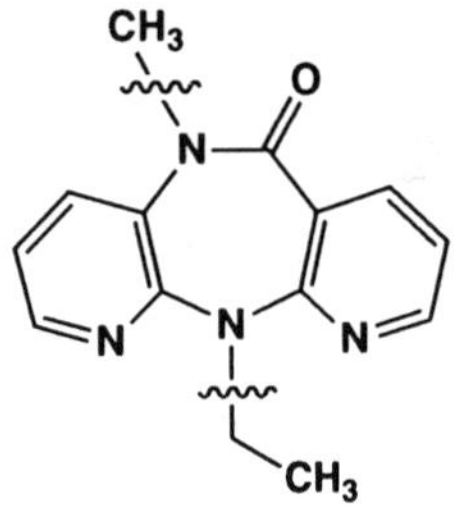

BI-RH-397

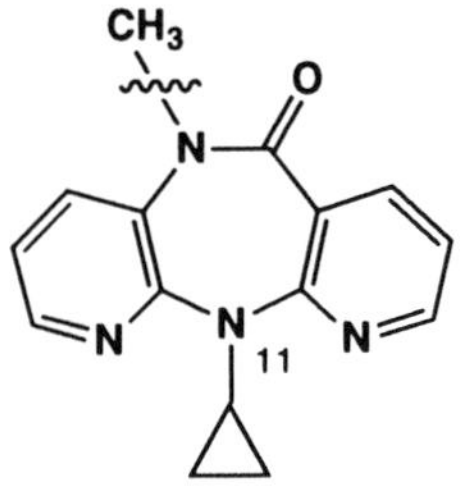

BI-RG-580

FIGURE 6. Metabolic *N*-dealkylation. The cyclopropyl group was introduced to prevent metabolic *N*-11-dealkylation.

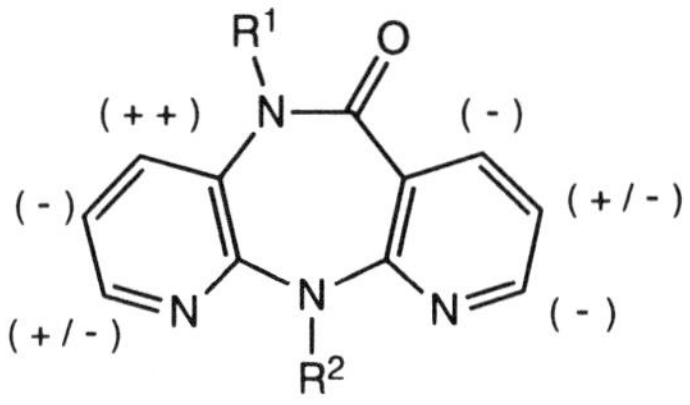

FIGURE 7. Summary of preliminary SAR of substituents on pyridine rings.

Compound Selection

It appeared that the analogue BI-RH-414 and the cyclopropyl derivative BI-RG-587 (nevirapine) both had the potency requirements originally sought. Both compounds showed good antiviral activities in the cell culture assay with IC_{50} values of 30–40 nM reproducibly. No cytotoxic effects of the compounds in cells was detected up to high concentrations (>300 μM) offering a high therapeutic index. Receptor binding assays indicated that the drugs possessed no significant activity, with the exception that BI-RH-414 exhibited binding to the rat peripheral benzo-diazepine receptor (K_i = 330 nM) whereas BI-RG-587 (nevirapine) was inactive up to micromolar concentrations. This result was difficult to interpret, however, without a better understanding of the pharmacological consequences of the peripheral receptor. Without further distinguishing characteristics in vitro, a detailed "head-to-head" comparison of both drugs was undertaken (Figure 9).

Oral dosing of BI-RH-414 and BI-RG-587 (nevirapine) revealed fairly similar profiles. The *N*-ethyl analogue BI-RH-414 was slightly more soluble than the corresponding cyclopropyl derivative. In both cases good blood levels of the parent drug were observed in rats, dogs, and cynomolgus monkeys. Nevertheless, BI-RG-587 (nevirapine) demon-

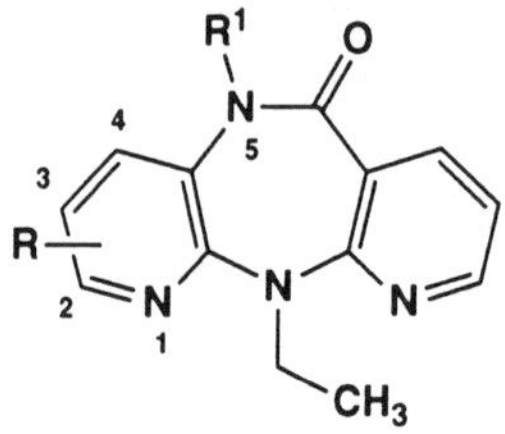

	R	R^1	HIV-1 RT IC_{50} (nM)
BI-RH-397	H	CH_3	125
	2-CH_3	CH_3	170
	3-CH_3	CH_3	760
	4-CH_3	CH_3	865
	H	H	475
	2-CH_3	H	1000
	3-CH_3	H	865
BI-RH-414	4-CH_3	H	35

FIGURE 8. Substitution in the 4-position, in combination with an unsubstituted amide ($R^1 = H$), greatly enhances inhibitory activity.

strated a better absorption profile by a small but consistent margin. This was most notable in the monkey. The anticipated N-deethylation metabolism for BI-RH-414 was not observed. Both drugs exhibited the same metabolic fate whereby the 4-methyl group was oxidized to give the 4-hydroxymethyl product by P-450 enzymes as the major metabolite, but this occurred to only a minor extent. Other trace oxidative metabolites were also observed and determined to be ring hydroxylated products, but this occurred to an extent less than 1%.

The differences between the two drug candidates were still rather minor, and it was deemed necessary to assess the safety of both compounds in the hope of bringing out any toxicity problems with the dipyridodiazepinone series as well as with each compound specifically. A 2-week "mini-safety" study was conducted using both rising dose and multiple dose protocols in both rats and dogs. The compounds were

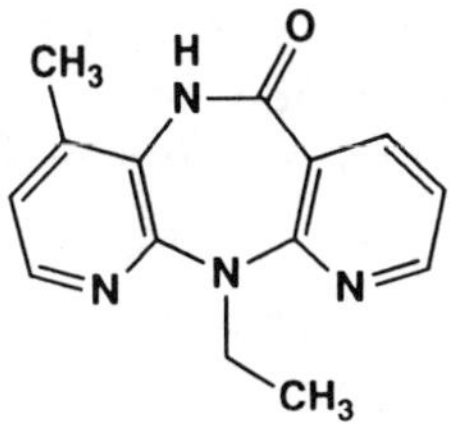

BI-RH-414

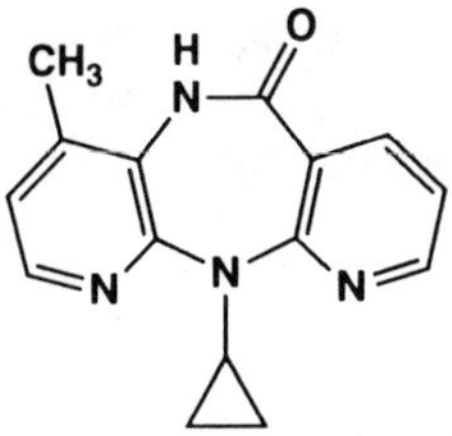

BI-RG-587
(Nevirapine)

FIGURE 9. Potential clinical candidates.

indistinguishable in their profiles. No significant toxicity or safety problem was uncovered.

Since the comparison of both lead structures was so similar in profile, the research team studied the data, trying to make a compelling argument to propose one compound for clinical development. In the end, two factors favored BI-RG-587 (nevirapine) as the consensus choice: (1) the binding of BI-RH-414 to the peripheral benzodiazepine receptor was of dubious significance, yet this activity was best avoided if possible; (2) BI-RG-587 (nevirapine) showed a better overall absorption profile in our animal studies, and bioavailability was certainly a property that should be maximized (Figure 10).

	BI-RH-414	BI-RG-587
		(Nevirapine)
Potency		
enzyme assay	35 nM	84 nM
cell culture assay	30 nM	40 nM
Solubility	0.17 mg/mL	0.1 mg/mL
Bioavailability	+	++
Toxicology	Not significant	Not significant

FIGURE 10. Selection of clinical candidate.

In February 1990, 15 months after the identification of the first lead RT inhibitor, management accepted the recommendation to develop BI-RG-587 (nevirapine) and a formal preclinical phase of the program was launched.

Mechanism of Action of Nevirapine

Accompanying the lead optimization and preclinical activities arose the need to understand the mechanism of action of nevirapine at a molecular level. The enzymatic screening assay reliably indicated that the tricyclic inhibitors were effective polymerase inhibitors. Furthermore, a good structure–activity correlation between inhibition of the enzyme and inhibition of viral replication in cell culture of a series of compounds strongly implied that the antiviral activity was indeed a result of the blockade of reverse transcription. Nevertheless, this was only circumstantial evidence that nevirapine was behaving as a true enzyme inhibitor through a discreet binding mechanism. A group of scientists concerned themselves with defining the precise mode of action of nevirapine. A multifaceted approach was adopted comprising enzyme kinetics, photoaffinity labeling, and spectroscopy. The following paragraphs will outline the key experiments elucidating the mechanism of action of nevirapine

Functional reverse transcriptase of HIV-1 is a heterodimeric protein consisting of 66- and 51-kD subunits. The active enzyme arises from proteolytic cleavage of the p66 homodimer by either cellular proteases or the virus's own protease to produce the p66/p51 heterodimer. The polymerase activity has been proposed to reside in the N-terminal portion of the p66 protein in the heterodimer. Ribonuclease H activity is associated with the C-terminal position of p66. For the purposes of detailed mechanistic studies, HIV-1 RT was purified to homogeneity and used for all the subsequent studies.

Steady-state enzyme kinetics were performed to determine the nature of the inhibitory action of nevirapine. Early on, homopolymer template-primer substrates were employed, but these data were also corroborated using a synthetic 59-base RNA heteropolymer and a corresponding 20-base DNA primer. Irrespective of the substrates employed, Lineweaver–Burke analysis indicated that nevirapine inhibits in a noncompetitive manner with respect to deoxynucleoside triphosphate or template-primer. This is depicted by the double reciprocal plots (velocity vs. substrate concentrations) (Figure 11). Control experiments also indicated that nevirapine does not intercalate duplex DNA or RNA–DNA heteroduplex.

The next phase of the investigation involved the design of a photo-

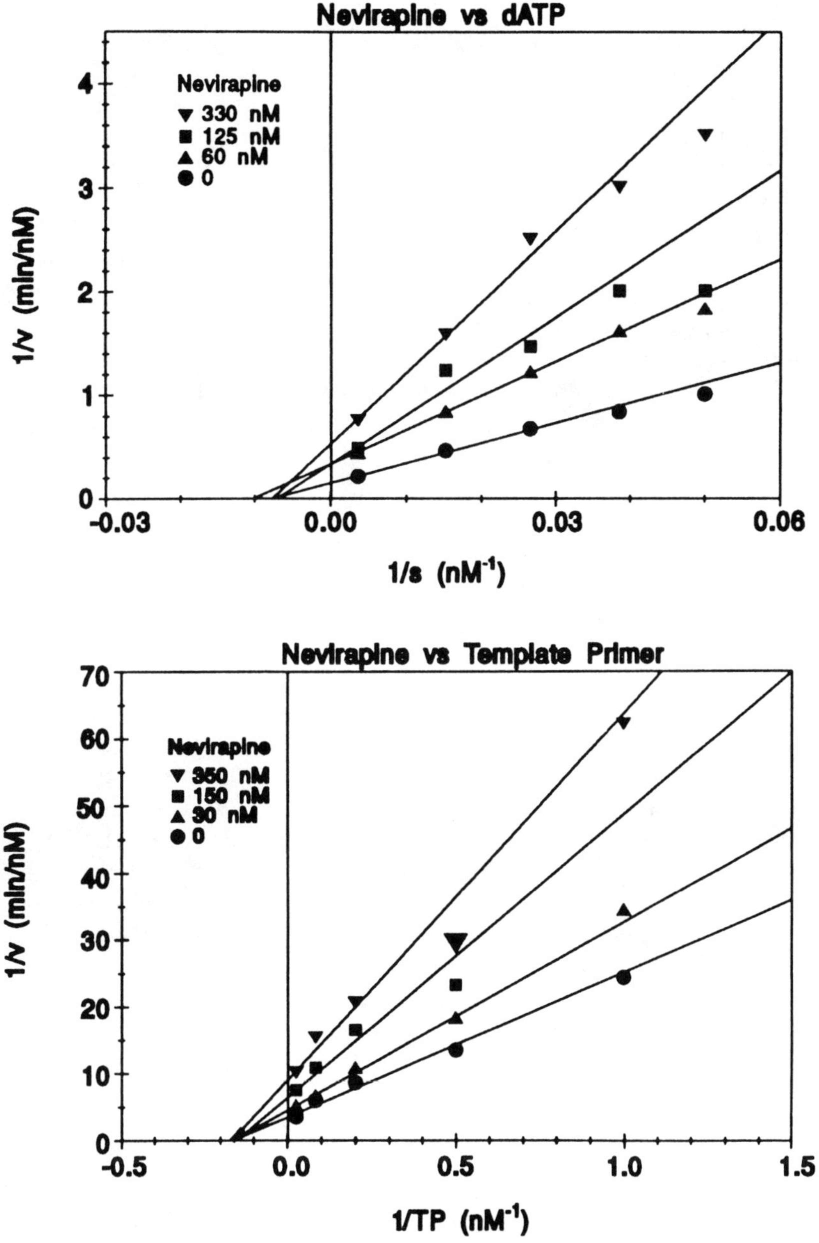

FIGURE 11. Steady-state kinetic analysis of reverse transcriptase inhibition of nevirapine using an authentic heteropolymer template.

affinity probe with the aim of covalently linking a structurally related analogue at the nevirapine binding site. Furthermore, the probe was also radiolabeled using tritium to facilitate easy detection and sequencing of the amino acids in the binding site. The chemists prepared an azide analogue (BI-RJ-70) with a tritium label in the C ring (Figure 12). In the absence of light this compound behaves like a reversible inhibitor (IC_{50} = 140 nM). However, upon exposure to ultraviolet (UV) light, the azide absorbs energy and releases nitrogen, producing a reactive nitrene intermediate that in the presence of the enzyme can covalently attach to proximal amino acid residues. Photolabeling with BI-RJ-70 was shown to be specific, since the enzyme could be protected from covalent labeling in a dose-dependent manner by adding nevirapine (Figure 13). It was also shown that the addition of nucleoside triphosphate or template-primer substrates did not protect the enzyme from labeling. This corroborated the results of the enzyme kinetic experiments, which indicated noncompetitive binding of the drug. Thus the nevirapine binding site is distinct from the substrate binding site, and the effect of drug binding is to lower the velocity at which the enzyme polymerizes DNA.

The photolabeling of the enzyme with BI-RJ-70 was time dependent and led to irreversible inactivation of the enzyme. The stoichiometry of labeling was determined (Figure 14), and it was found that one molecule of BI-RJ-70 per molecule of enzyme was sufficient to completely inhibit polymerase activity. The labeled enzyme was recovered, and SDS–PAGE [sodium decyl sulfate–poly(acrylamide) gel electrophoresis] analysis revealed that only the p66 protein was labeled during the photoinactivation phase. Continued irradiation and further addition of BI-RJ-70 eventually led to nonspecific labeling of p66 and p51 proteins. The specificity of the drug binding to RT was further demonstrated by performing photoaffinity labeling experiments in the presence of cellular proteins (derived from peripheral blood lymphocytes). Autoradiographs of the protein gels once again demonstrated the specific labeling of the RT p66 protein (Wu et al., 1991).

The RT enzyme bound with the radiolabeled BI-RJ-70 on p66 was subjected to enzymatic digestion with trypsin. This produced a mixture of about 75 peptide fragments, which by HPLC (high-pressure liquid chromotography) analysis could be separated and purified. Detection either with a UV detector or a radioactivity detector permitted the isolation of a peptide containing the radiolabeled BI-RJ-70 adduct. The peptide fragment was sequenced, and the amino acid identities were determined. The peptide corresponded to the sequence defined by amino acids 174–199 of p66, and the photoaffinity probe was covalently bound to tyrosines 181 or 188 (Figure 15). This part of the enzyme molecule is predicted to be proximal to the active site based on the

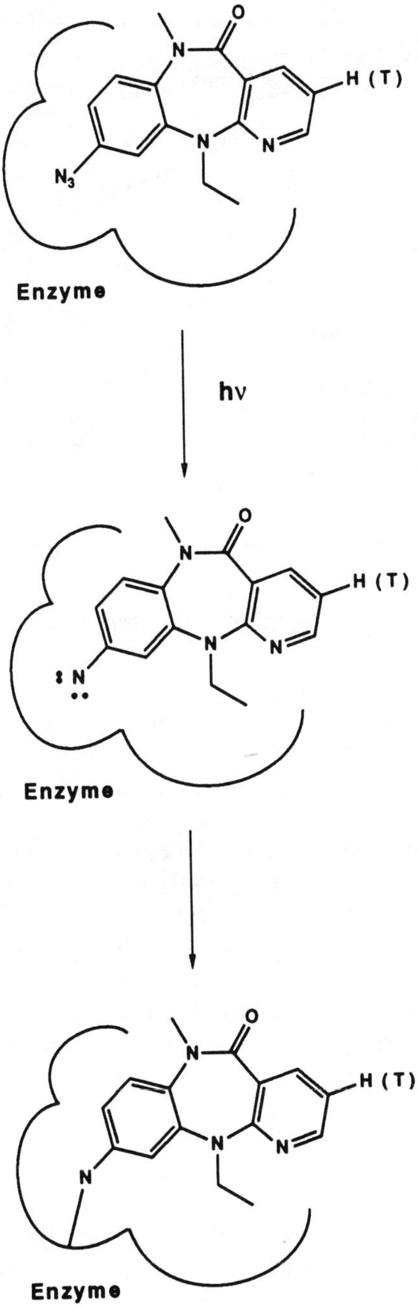

FIGURE 12. Photoaffinity labeling of reverse transcriptase.

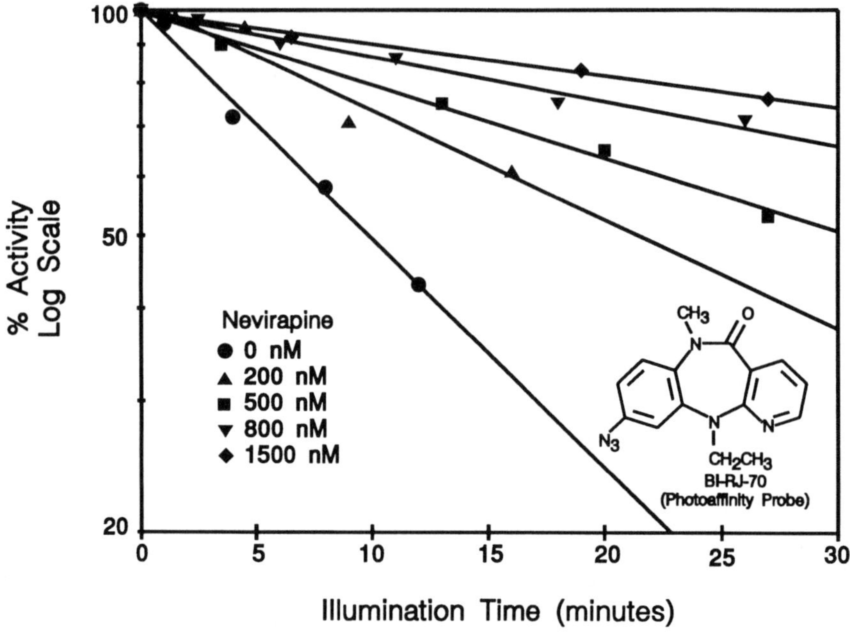

FIGURE 13. Time-dependent inhibition of HIV-1 reverse transcriptase by the photoaffinity probe in the presence of nevirapine.

highly conserved 183–187 sequence in polymerases in general (Cohen et al., 1991).

Further confirmation that nevirapine binds in the region of tyrosines 181 and 188 was obtained through site-directed mutagenesis (Figure 16). Specific substitution of tyrosine 181 to isoleucine (Tyr181Ile), the amino acid in HIV-2 RT at the analogous position, reduced nevirapine's inhibitory activity by 3 orders of magnitude. Similar results were obtained with Tyr188Leu mutations (Shih et al., 1991). Further spectroscopic studies, fluorescence transfer techniques, and circular dichroism measurements of the enzyme in the presence and absence of nevirapine analogues demonstrated specific binding of drug to the enzyme molecule. Furthermore, when the tyrosine mutants were examined, a commensurate loss of binding affinity could be correlated (C. A. Grygon, D. J. Greenwood, R. C. Cousins, and J. Stevenson, unpublished data, 1991). Taken together the data were highly consistent and provided a mechanistic model for the inhibition by nevirapine of HIV-1 RT.

Once the mechanism studies were going on within Boehringer Ingelheim Pharmaceuticals, several of us decided to make an overture to Professor Tom Steitz at Yale University, who is considered an authority in the structural biochemistry of polymerases and had successfully

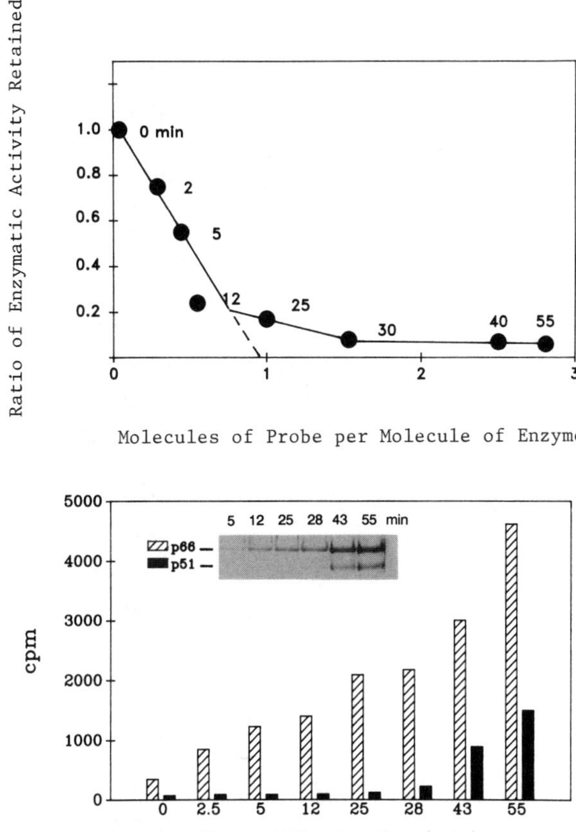

FIGURE 14. Stoichiometry and localization of photoaffinity labeling.

solved the crystal structure of the Klenow fragment of *E. coli* polymerase. RT had in fact been crystallized by several groups, but the crystals obtained defracted only at poor resolution (7–10 Å). We reasoned that perhaps RT bound with nevirapine might provide a different crystal form and offer a better resolved structure. We provided some funding, and Dr. Steitz agreed to take in one of our biochemists to proceed with cocrystallization studies. In retrospect this was a naive venture, but one that certainly paid off. After 6 months of fruitless effort, success was achieved with a 3.5-Å resolved cocrystal of RT and nevirapine. An intense period ensued wherein the Yale group worked tirelessly on solving the three-dimensional (3-D) structure. Approximately 1 year after the first crystals were grown, we were treated to the first 3-D view of nevirapine bound to RT in the predicted hydrophobic pocket containing Tyr181 and 188 (Kohlstaedt et al., 1992). This landmark event triggered new activities that remain ongoing to further understand how

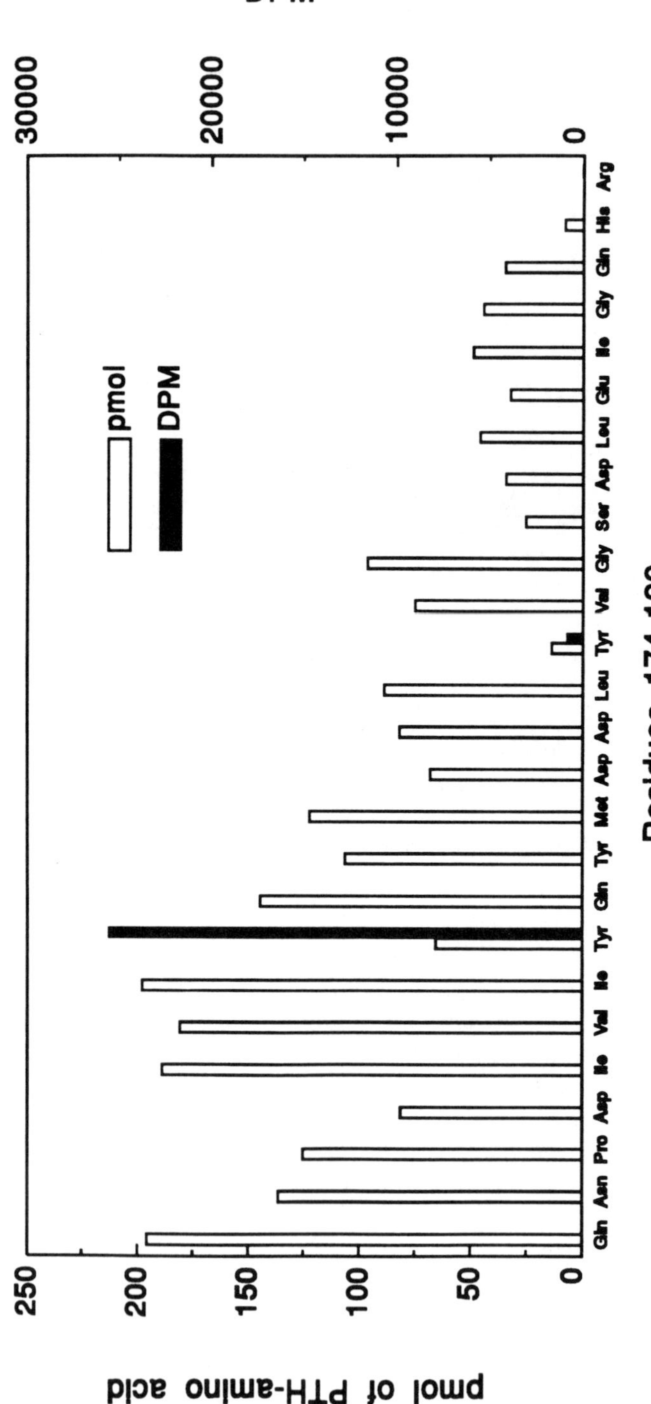

FIGURE 15. Sequence of HIV-1 RT tryptic fragment 24.4 labeled with BI-RJ-70.

64

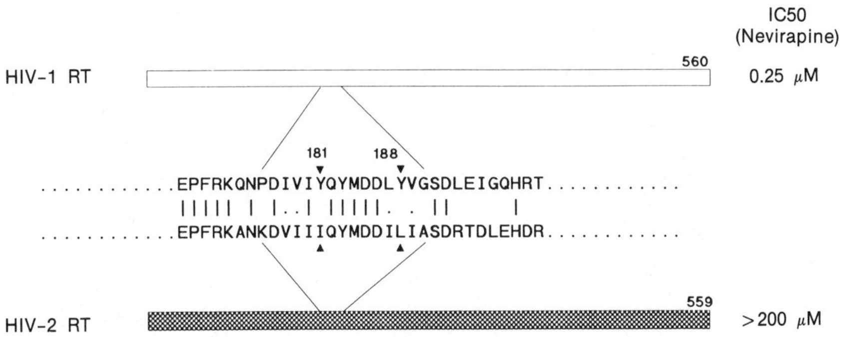

FIGURE 16. Mutational sensitivity to nevirapine. Although HIV-2 RT has 61% amino acid sequence identity to HIV-1 RT, it is not sensitive to nevirapine.

nevirapine works and of course to design an improved second-generation RT inhibitor.

Effects of Nevirapine on Virus Replication: Deeper Evaluation

Once nevirapine was chosen for development, a few key issues had to be studied. These included effects of nevirapine on the immune system and especially bone marrow, as well as effects of nevirapine on virus replication, detailed studies on AZT-resistant strains, clinical isolates, and synergy with AZT and other nucleosides. Inherent in these latter studies was the possibility of developing resistance to nevirapine itself.

While it was clear that nevirapine inhibited virus replication as measured by the inhibition of syncytia and p24 production in culture using laboratory strains, it was necessary to study the effects of nevirapine on clinical isolates as well. Clinical isolates were obtained from Dr. Gary Wormser at New York Medical College and were tested in syncytia and p24 assays by Drs. John Sullivan and Richard Koup at the University of Massachusetts Medical School. The results of four of these isolates are shown in Figure 17. Nevirapine inhibited virus replication in all isolates. This has since been confirmed and extended to other isolates. In addition a nonisotopic in situ hybridization experiment on one of the isolates showed that in HIV-1 infected fresh normal peripheral blood mononuclear cells, 69/830 cells were positive for HIV-1 RNA whereas only 5/93,000 cells were positive for HIV-1 RNA when the cultures were treated with nevirapine (Merluzzi et al., 1990).

Although it was clear to most that nevirapine inhibited HIV-1 replication specifically, it was suggested to us that we ought to confirm the

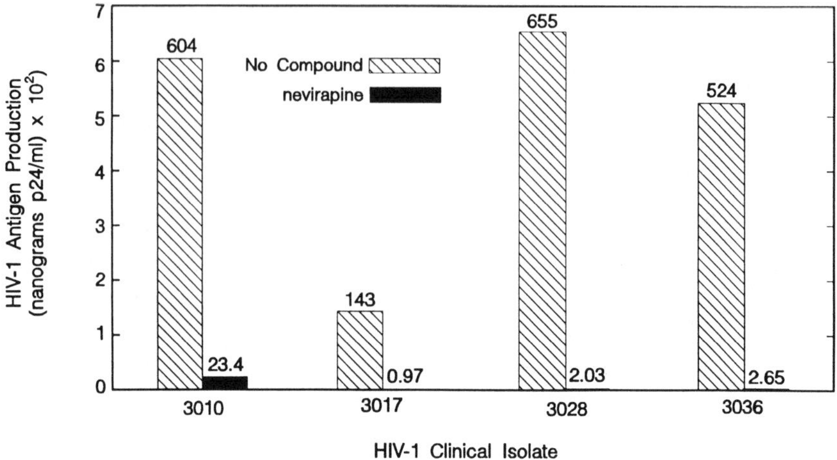

FIGURE 17. Effect of nevirapine on HIV-1 clinical isolates.

activity in a separate laboratory using different systems. Prior to this point, the replication of HIV-1 and its inhibition was accomplished using a syncytia assay and a p24 assay in c8166 lymphoblastoid T-cells (Merluzzi et al., 1990; Koup et al., 1991). Dr. Douglas Richman kindly provided us with results using nevirapine in a plaque assay using CD4-transfected HeLa cells. In this assay, nevirapine inhibited virus replication with an IC_{50} of 15 nM while having no effect on HIV-2 and other RNA viruses. A comparison of the activities of nevirapine and AZT is shown in Table 1 (also Richman et al., 1991a). Once the antiviral activity of nevirapine was confirmed, it was necessary to study the effect of this compound on AZT-resistant strains and to study its use together with AZT for additive or synergistic effects. Since it had been shown that nevirapine binds to a site other than the active site of RT and acts as an allosteric noncompetitive inhibitor of RT, we suspected that nevirapine would inhibit AZT-resistant strains and the combination of two agents might inhibit HIV-1 in a synergistic manner. As can be seen in Figure 18, nevirapine inhibits AZT-resistant strains. Furthermore com-

TABLE 1. Reduction by Nevirapine and AZT of Plaque Formation by HIV-1 and HIV-2

Virus	Nevirapine (nM)			AZT (nM)		
	IC_{50}	IC_{90}	IC_{95}	IC_{50}	IC_{90}	IC_{95}
HIV-1	16	450	710	30	710	2,000
HIV-2	>3,200	–	–	112	>3,200	–

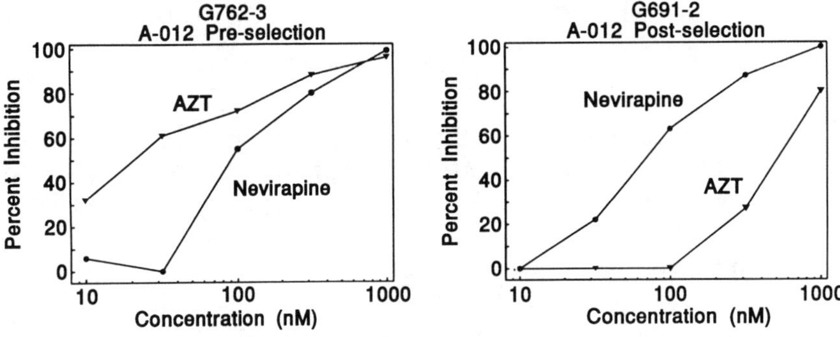

FIGURE 18. Effect of nevirapine on AZT-resistant HIV-1 isolates.

bination experiments of nevirapine with AZT demonstrated synergistic inhibition of HIV-1 (Richman et al., 1991a). Equally important, however, was the fact that nevirapine and AZT were not synergistic for toxicity either against uninfected T-cell lines (unpublished data) or against bone marrow progenitors (Moore and Merluzzi, 1991). As expected, viral resistance to nevirapine has recently been demonstrated in vitro, and together with previously mentioned RT mutagenesis data (Shih et al., 1991), resistance occurs at the 181 *aa* binding site (Richman et al., 1991b).

Immunology and Bone Marrow Progenitors

Early studies indicated that nevirapine had no suppressive effect on either the humoral or cellular arms of the immune response. Nevirapine did not suppress a delayed hypersensitivity response in rats to methylated bovine serum albumin (BSA) and likewise had little or no effect on the antibody response in mice to sheep red blood cells. It should be kept in mind that the doses of nevirapine used in these studies were 2 orders of magnitude higher than what we considered to be a therapeutic dose based on early pharmacokinetics. Although the immune response did not seem to be appreciably affected by nevirapine, its effects on bone marrow progenitors had to be tested in light of the effects of AZT on bone marrow cell cultures and the results of bone marrow depression. Dr. Malcolm Moore at Sloan-Kettering Cancer Center kindly tested nevirapine and AZT and other analogues in a coded fashion on erythroid colonies and mixed colonies of progenitors of myeloid, erythroid, and megakaryocytes. In all instances, nevirapine did not suppress bone marrow colony formation except at very high concentrations, whereas AZT was suppressive at all concentrations tested (Moore and Merluzzi, 1991).

Investigational New Drug Application (IND), Clinical Trials, and Disclosures

It was clear that by mid-1990 that the basic enzymology data, virus replication data, "mini-safety" results, and pharmacokinetics supported the development of nevirapine as a drug candidate. These data were presented to the National Institutes of Health (NIH), and design of long-term toxicology and clinical trials began soon thereafter. A development team handled issues concerning scaleup, production, formulation, stability, and deeper pharmacology/toxicology of nevirapine. The goal of the team was to file an IND in the fourth quarter of 1990 with enough safety data to proceed from a single-dose safety trial to multiple dosing in early 1991. Both the NIH and the Food and Drug Administration (FDA) as well as our scientific and clinical collaborators were instrumental in formulating a clinical plan with our own clinical investigators. In our desire to move quickly with nevirapine, we met our goals and filed an IND in December 1990 and began clinical testing of nevirapine in January 1991. This filing was 6 months to 1 year sooner than a normally paced development project. The first publication on nevirapine occurred in an abstract in October 1990 at the annual ICAAC (Int. Conf. of Antimicrobial Agents & Chemotherapy) conference in Atlanta, followed by a publication in *Science* in December 1990. The publication received noteworthy exposure and raised the interests of the scientific community to the idea of nonnucleoside inhibitors of RT. The initial publication of a nonnucleoside inhibitor of RT in March 1990 (Pauwels et al., 1990) in *Nature* solidified in our minds and others that this type of inhibitor was both real and worthy of expedited development.

Phase I studies were carried out in AIDS patients, and the pharmacokinetic and safety profiles for nevirapine were very encouraging. The drug was well tolerated and found to have a mean half-life of about 30 hours, which was ideal for once-a-day dosing. Single- and multiple-dose studies of nevirapine alone and in combination with AZT were also conducted, as well as the testing of a suspension formulation in pediatric AIDS patients. Studies were conducted under the auspices of the AIDS Clinical Trials Group (ACTG) in three centers including the University of Massachusetts, the University of Minnesota, and the University of California at San Diego. Multiple-dose trials seeking to demonstrate the antiviral effect of nevirapine were undertaken. Starting at low dose (12.5 mg/day) and escalating the dose (up to 400 mg/day), we saw a clear dose-responsive reduction of p24 antigen at a period of 16 weeks of therapy.

We also experienced our first clinical setback. We soon discovered that viral isolates with reduced sensitivity to nevirapine could be obtained within 4 weeks of nevirapine therapy. The emergence of this resistence was prolonged in time at higher doses. We soon decided to abandon low-dose treatment and placed all willing patients on a high-dose regimen (400 mg/day). Additionally, we resolved to study combination therapy with the approved nucleoside antivirals.

Subsequent to our discovery and that of Pauwels et al. (1990), Goldman et al. (1991) have also described a nonnucleoside inhibitor of RT. All three nonnucleoside inhibitors (Merluzzi et al., 1990; Pauwels et al., 1990; Goldman et al., 1991; also see Chapters 4 and 5 in this volume) bind to the same pocket on RT and have similar profiles for inhibiting HIV-1 replication. All three compounds were discovered independently and during the same time frame. The outcome of these nonnucleoside inhibitors as antivirals for HIV-1 should be known over the next few years.

Acknowledgment and Afterword

The discovery and development of nevirapine was a very exciting period for Boehringer Ingelheim. Many individuals, though here unnamed, provided important contributions which in the aggregate made this project succeed in record time. The interplay of disciplines was historic in the company and provided a model for future drug discovery.

References

Cohen, KA, Hopkins J, Ingraham R, Pargellis C, Wu JC, Palladino DEH, Kinkade P, Warrent TC, Rogers S, Adams J, Farina PR, Grob P (1991): Characterization of the binding site for BI-RG-587, a non-nucleoside inhibitor of HIV-1 reverse transcriptase. *J Biol Chem* 266:14670–14674

Goldman ME, Nunberg JH, O'Brien JA, Quintero JC, Schleif WA, Freund KF, Gaul SL, Saari WS, Wai JS, Hoffman JM, Anderson PS, Hupe DJ, Emini EA, Stern AM (1991): Pyridinone derivatives: Specific human immunodeficiency virus type 1 reverse transcriptase inhibitors with antiviral activity. *Proc Natl Acad Sci USA* 88:6863–6867

Kohlstaedt LA, Wang J, Friedman JM, Rice PA, Steitz TA (1992): Crystal structure at 3.5 A resolution of HIV-reverse transcriptase complexed with an inhibitor. *Science* 25:1782–1790

Koup RA, Merluzzi VJ, Hargrave KD, Adams J, Grozinger K, Eckner RJ, Sullivan JL (1991): Inhibition of human immunodeficiency virus type 1 (HIV-1) replication by the dipyridodiazepinone BI-RG-587. *J Infect Dis* 163:966–970

Merluzzi VJ, Hargrave KD, Labadia M, Grozinger K, Skoog M, Wu JC, Shih C-K, Eckner K, Hattox S, Adams J, Rosenthal AS, Faanes R, Eckner RJ, Koup RA, Sullivan JL (1990): Inhibition of HIV-1 replication by a nonnucleoside reverse transcriptase inhibitor. *Science* 250:1411–1413

Moore MAS, Merluzzi VJ (1991): Comparison of a non-nucleoside HIV-1 reverse transcriptase inhibitor and zidovudine on human bone marrow progenitors. In preparation.

Pauwels R, Andries K, Desmyter J, Schols D, Kukla MJ, Bresline HJ, Raeymaeckers A, Van Gelder J, Woestenborghs R, Heykants J, Schellekens K, Janssen MAC, De Clerq E, Janssen PAJ (1990): Potent and selective inhibition of HIV-1 replication in vitro by a novel series of TIBO derivatives. *Nature (London)* 343:470–474

Richman D, Rosenthal AS, Skoog M, Eckner RJ, Chou T-C, Sabo JP, Merluzzi VJ (1991a): BI-RG-587 is active against zidovudine-resistant human immunodeficiency virus type-1 and synergistic with zidovudine. *Antimicrob Agents Chemother* 35:305–308

Richman D, Shih C-K, Lowy I, Rose J, Prodanovich P, Goff S, Griffin J (1991b): HIV-1 mutants resistant to non-nucleoside inhibitors of reverse transcriptase arise in tissue culture. *Proc Natl Acad Sci USA,* 88:11241–11245

Shih C-K, Rose JM, Hansen GL, Wu JC, Bacolla A, Griffin JA (1991): Chimeric human immunodeficiency virus type 1/type 2 reverse transcriptases display reversed sensitivity to nonnucleoside analog inhibitors. *Proc Natl Acad Sci USA* 88:9878–9882

Wu JC, Warren TC, Adams J, Proudfoot J, Skiles J, Raghavan P, Perry C, Potocki I, Farina PR, Grob PM (1991): A novel dipyridodiazepinone inhibitor of HIV-1 reverse transcriptase acts through a nonsubstrate binding site. *Biochemistry* 30:2022–2026

4

Discovery of TIBO, a New Family of HIV-1-Specific Reverse Transcriptase Inhibitors

Rudi Pauwels

Introduction

It was with more than usual interest that we learned about the essential role of a human retrovirus, HIV (human immunodeficiency virus), in the pathogenesis of AIDS (acquired immune deficiency syndrome). Since the early 1970s the group of Prof. Erik De Clercq at the Rega Institute for Medical Research at the Catholic University in Leuven (or Louvain, Belgium) had been focusing on antiviral chemotherapy. The wide range of viruses that were studied also included murine retroviruses, for which in vitro and in vivo antiviral models were being developed. Such studies indicated that, at least theoretically, several stages in the HIV replicative cycle might be vulnerable to chemotherapeutic intervention: adsorption and interaction with the viral receptor; penetration/uncoating; reverse transcription (and the accompanying degradation of the viral RNA template); integration of the DNA into the host genome; expression (transcription and translation) of the viral genes, and the regulation of this expression; processing of the viral precursor proteins through myristylation, proteolysis, and glycosylation; virus assembly; and budding of newly formed virions.

After they had demonstrated its efficacy on HIV replication in cell culture (Mitsuya et al., 1984), the group of Drs. S. Broder, R. Yarchoan, and H. Mitsuya at the National Institutes of Health (NIH) administered suramin, whose antiretroviral properties were reported in 1979 by the Rega group (De Clercq, 1979), to AIDS patients (Broder et al., 1985). It was expected that, if one could block HIV replication efficiently, pathogenesis of AIDS and its clinical manifestations also would be suppressed. This hypothesis would bear fruit with the advent of

The Search for Antiviral Drugs
Julian Adams and Vincent J. Merluzzi, Editors
© 1993 Birkhäuser Boston

3′-azidothymidine (AZT, zidovudine, Retrovir®), the first drug shown to benefit AIDS patients (Yarchoan et al., 1986; Fischl et al., 1987). Unfortunately, it would soon become clear that AZT was not an ideal drug, as side effects, including myelotoxicity, were associated with prolonged drug treatment. Suramin and later AZT turned out to be typical instances of how the first generation of anti-HIV agents were discovered: compounds were selected for in vitro anti-HIV evaluation on the basis of their activity against other viruses, in particular retroviruses.

Start of the Anti-HIV Program

Under the impulse of Prof. Erik De Clercq and with the Suramin episode in the background, I started in 1985 by setting up a laboratory that would focus on the discovery and characterization of anti-HIV agents. This laboratory would be located within the clinical virology department of Prof. Jan Desmyter, who had just obtained several HIV-containing vials from Dr. Luc Montagnier of the Pasteur Institute in Paris. We decided first to develop an anti-HIV cell culture assay (see below) to test agents which had shown previously interesting antiretroviral activities. In addition we wanted to evaluate new 2′,3′-dideoxynucleoside analogues synthesized by Prof. Piet Herdewijn and Dr. Arthur Van Aerschot from our institute. Together with Dr. Masanori Baba, who joined us in the summer of 1986, we evaluated and described several 2′,3′-dideoxynucleoside analogues with potent and selective anti-HIV activity (Baba et al., 1987). Screening of a large variety of phosphonate derivatives synthesized by Dr. A. Holy (Czechoslovakia) also led to the discovery of the first acyclic nucleoside analogue (9-(2-phosphonylmethoxyalkyl)adenine, PMEA) with selective anti-HIV properties (Pauwels et al., 1988b).

Meanwhile, Dr. Koen Andries from the Janssen Research Foundation (JRF) was heading a successful program on antivirals against rhinoviruses in Beerse, Belgium (see Chapter 8 by Andries in this volume). I had met him several years earlier during an antiviral meeting in Rotterdam, after which we kept each other informed about the progress in our respective research areas. Early 1987, Dr. Paul Janssen, founder and chairman of Janssen Pharmaceutica, took a firm interest in the anti-HIV chemotherapy area. Based on his research philosophy, which had led various therapeutic areas to major breakthroughs, and success in the search for antirhinovirus compounds, he was convinced that it would be possible to find new inhibitors of HIV replication if only the right way of searching for these agents was applied, i.e., rational screening with a whole-cell assay. He visited our new anti-HIV laboratory in Leuven and discussed the status of our research. As it was immediately clear that both groups had a common view on the development of new anti-HIV agents, the Janssen Research Foundation and the Rega Institute started a formal collaboration: Dr. Andries would coordinate the program in Beerse, while I would organize the virological studies in Leuven. I started screening the first compounds of the JRF compound repository in our laboratory in July 1987 together with Hilde Azijn, a talented technician who had just finished her studies. Soon we were joined by Dominique Schols, a biologist who would specialize in studies using Fluorescence-Activated Cell Sorter (FACS) analysis and primary cells such as peripheral blood lymphocytes and monocytes/ macrophages.

Rational Screening

The goal of our collaboration was to develop totally new types of anti-HIV drugs, i.e., compounds with novel nonnucleoside chemical structures, having good oral bioavailabilities and pharmacokinetic profiles. We thought that, if we could block HIV replication in vivo, further impairment of the immune status in HIV-infected patients would slow down and possibly stop. To this end we first needed to find inhibitors of HIV replication in vitro that were sufficiently potent and selective. The antiviral target of the drug as such was of secondary importance. After all, we would only consider a compound valuable if it proved active and selective in a cell culture assay. We hoped of course that we would discover new targets in the HIV life cycle as well. Dr. Janssen and his group of chemists had synthesized about 80,000 new chemical entities (NCEs) since 1953, known at the JRF as R-numbers. Of these NCEs, 69 have been marketed, of which 48 are being used for human application in several fields of medical therapy. However, this unique compound repository also contained a number of biologically active compounds and a large set of substances with no known pharmacological or toxicological activity. The fact that all the compounds ever synthesized at the JRF had been evaluated in a number of standard pharmacological and toxicological in vitro and in vivo assays and that all these compounds as well as their information were readily available contributed significantly to the success of the program.

Whereas many teams either were screening substances at random or were screening compounds which had been known for many years to interfere with one or another virus function, we used a selection of 600 compounds having no known pharmacological effect. This selection was made at the start of the antirhinovirus program in Beerse and included prototypes of different chemical series. However, it was already recognized at the time that "prototypes" is a relative term, and therefore this list would later be extended to 2000 compounds. The philosophy behind this strategy is that, by selecting from novel chemical structures, we increased our chances of finding novel types of interaction. This is the first component of the approach which one could describe as "rational screening." A schematic diagram of this rational drug screening strategy, as compared to rational drug design, is shown in Figure 1.

Not only the selection of the candidate molecules for screening but also the way in which the screening was performed was based on a number of defined choices. We were convinced that the screening assay had to be based on replicating virus in a virus–cell culture system. Measuring the effect of selected compounds at the cellular level minimized the risk of missing any interesting compounds that might have been overlooked if the assay were based on isolated target proteins only. Moreover, since the virus completed several replicative cycles, all possible targets, including those that were not yet identified, could be adequately assessed. Furthermore, the results of such an assay would tell us that the compound was active in a complex biological matrix and reflected cellular uptake, stability, and metabolism of the drug. Again, this meant that if an active molecule would be found, we only had to consider optimizing the anti-HIV potency. On the other hand, cell culture assays are traditionally considered to be tedious, longer in duration, and more susceptible to biological variation. In addition,

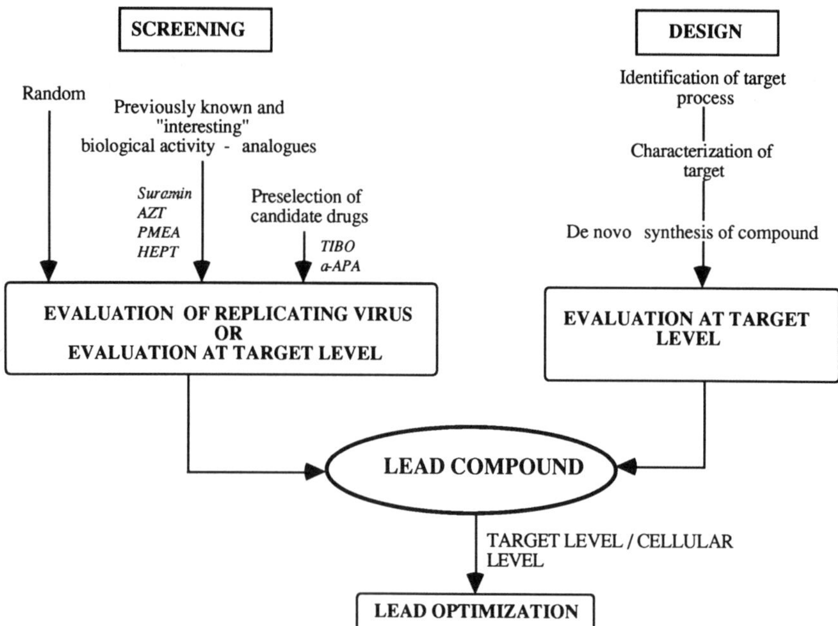

FIGURE 1. Schematic diagram of possible strategies for discovery of anti-HIV agents.

working with HIV presented a biosafety problem. Yet, our laboratory
was designed to do this type of work. We had about two years of HIV
experience, and the idea of working with this newly discovered virus,
for which antiviral drugs were badly needed, was more of a challenge
and a motivating factor than a threat to our new and young team.

The Screening Assay

An inherent consequence of the rational screening approach was that a
large number of compounds had to be evaluated. Therefore we decided
to follow a multistep procedure (Figure 2). A so-called primary assay
would first filter out the "interesting compounds," i.e., compounds
which inhibit HIV replication at concentrations that do not impair the
normal cellular functions of the host. Once we would identify such a

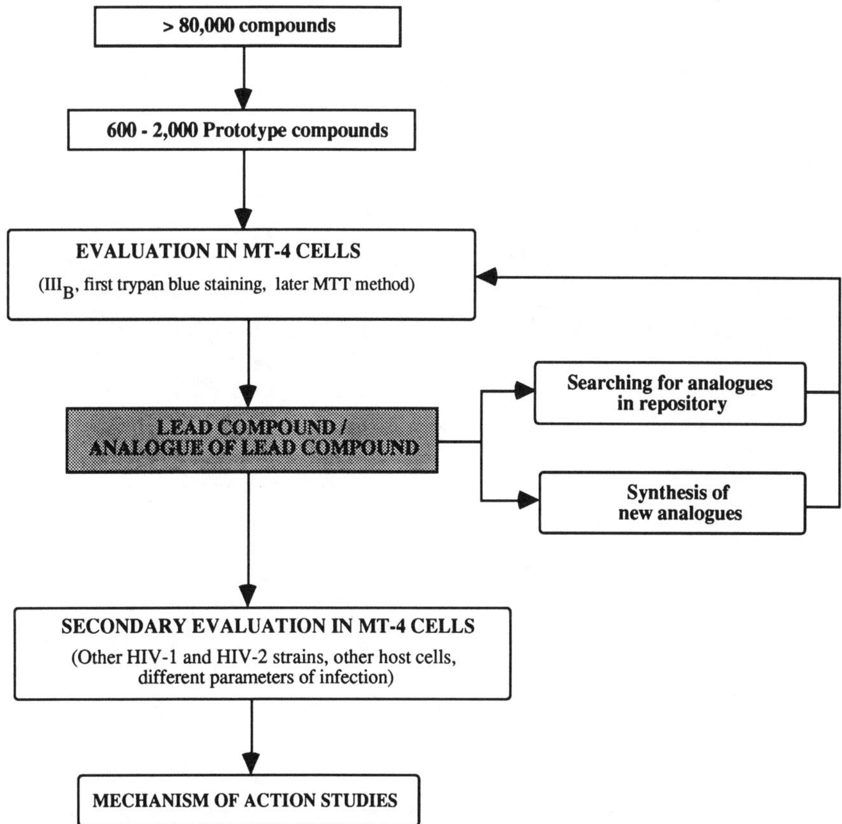

FIGURE 2. Flow diagram of virological studies.

"lead" compound, we would search the chemical library for related compounds and subject them to our primary assay. This would rapidly generate information on the anti-HIV structure–activity relationship (SAR). On the basis of this SAR information and/or the potency and selectivity of the lead compounds, it would be decided whether to initiate a new chemical synthesis program or not. At regular time intervals, more detailed virological studies were carried out with a most promising compound. This secondary evaluation phase included determination of the activity of the compounds in other host cells, measurement of the effects of the compound on different viral parameters, and investigation of the antiviral spectrum. In a third phase we would tackle the mechanism of action.

Mitsuya and co-workers (1984) described the first anti-HIV assay in the CD4+ T-cell line H9 (a clone of the better known HUT-78 line), in which viral replication was monitored by reverse transcriptase activity. However, most of their screening was apparently performed in ATH8 cells, an HTLV I–transformed CD4+ T-cell line. HIV infection of these cells led after about 7 days incubation to a cytopathic effect, loss of cell viability, and a characteristic change in cell pellet appearance. Since the cells were grown in large tubes, the cell pellet could be easily observed by the naked eye.

As the impairment of T4-cell function was considered to play a key role in the progression of the immunodeficiency, we also preferred to determine the anti-HIV properties of the compounds first in T-cell models. Although Mitsuya and co-workers had used tubes, most likely for safety reasons, I decided from the beginning to try to set up anti-HIV assays in 96-well microtiter plates because this format is easy to handle and has a large capacity. Moreover, several manufacturers had developed a range of tools for manipulating these plates. If the necessary precautions were taken, they could be handled safely as well. Prior to the start of our program we had already compared a variety of CD4+ T-cell lines for their susceptibility to HIV infection and their growth properties in 96-well microtiter plates. Among the cells were also the MT-4 and MT-2 cells which had been obtained from Dr. Luc Montagnier. These HTLV I–transformed CD4+ T-cell lines had been reported several months earlier by Prof. N. Yamamoto and co-workers (see Koyanagi et al., 1985) to be highly permissive for HIV infection. At the time the collaboration between the Rega group and the JRF group started we had developed an assay in 96-well plates using the MT-4 cells which was now selected as our primary assay. In this assay serial dilutions of compounds were incubated in 96-well microtiter plates for 5 days with MT-4 cells that had been either mock-infected or infected with HIV-1. During this incubation period the virus goes through several rounds of replication, eventually leading to a complete cytopathic effect (CPE) (Figure 3). Under these conditions all known and unknown steps of the replicative

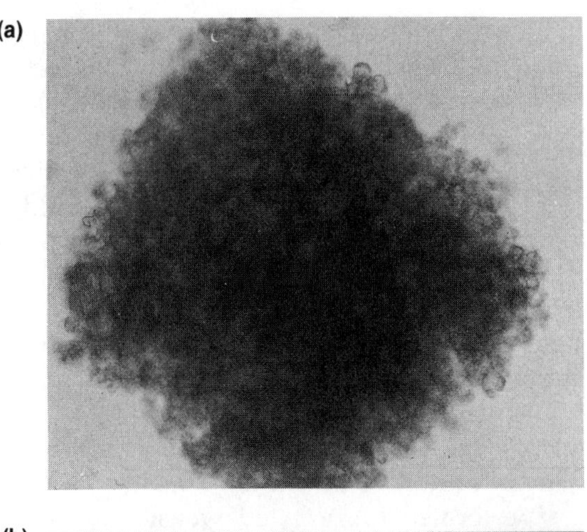

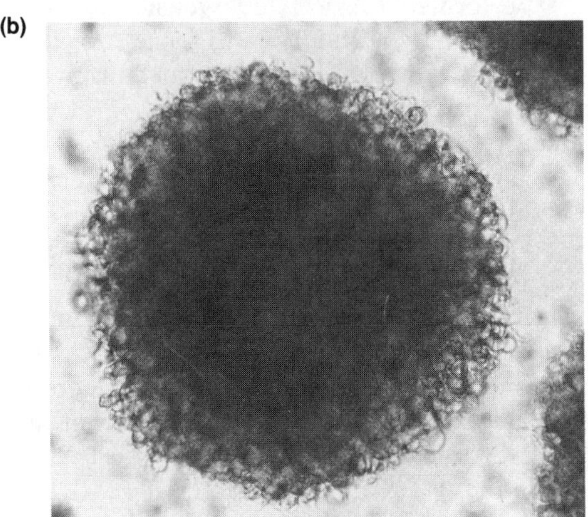

FIGURE 3. HIV-induced cytopathicity in MT-4 cells. HIV-infected (a) and mock-infected (b) MT-4 cells 5 days postinfection (a × 100; b, × 100).

cycle were adequately assessed. The HIV-induced CPE was determined by a microscopic viability staining method with trypan blue. We favored this type of analysis because both the anti-HIV activity as well as the cytotoxicity of the compounds could be determined in parallel and with the same methodology. In measuring other parameters such as reverse transcriptase activity and p24 HIV core protein production, one must bear in mind that a decrease of these values may be due to the compounds' anti-HIV activity and/or their cytotoxicity for the host cells and therefore, in the absence of cytotoxicity data, may be difficult to interpret. To keep the primary assay as simple as possible, the cell-seeding conditions were chosen so that the cells did not need to be subcultured and replenished with new compound during the incubation period. Prior to their addition to microtiter plates, the cells were infected with the prototype III_B/LAI strain at a multiplicity of infection that nearly destroyed all host cells after 5 days. On the one hand, lower multiplicities would lead to more false positives. On the other hand, lead compounds with a weak but selective anti-HIV activity would be difficult to detect if very high multiplicities were used. This readily available laboratory strain was used by several researchers in the field and hence was a good basis for comparison of our results. Later, when our own data base had expanded and other HIV-1 strains became available, we nevertheless kept the III_B strain as primary challenge virus for that same reason.

During the first months of our project all microtiter plates were first investigated microscopically to determine the concentration ranges of the anti-HIV activity and the cytotoxicity. Viability counts with trypan blue were then performed on selected wells (Pauwels et al., 1987). Comparison of these counts with HIV- and mock-infected controls were used to define the 50% antiviral effective concentration (IC_{50}) and the 50% cytotoxic concentration (CC_{50}). The ratio of CC_{50} to IC_{50} was defined as the selectivity index. The data were entered in the laboratory on so-called data sheets. The calculations were performed by a simple homemade spreadsheet program. Tables with the results of those calculations were regularly faxed to the JRF headquarters in Beerse, where they were entered manually in a data base.

As our initial screening capacity was only 10–20 compounds a week, it soon became clear that the trypan blue method was far too cumbersome and time consuming for the extensive numbers of compounds which needed to be examined. Therefore, significant automation was introduced in our procedure. The protective effects against the CPE, induced by a prototype HIV-1 strain III_B, as well as the host cell cytotoxicity of the compounds were quantified with the tetrazolium dye MTT [3-(4,5-dimethylthiazol-2-yl)-2,5-diphenyl-2H-tetrazolium bromide] (Pauwels et al., 1988a). This dye is taken up and metabolized in living cells only. It thereby generates a color which can be measured spectrophotometrically. Based on this principle, a new microtray assay procedure was developed that combined computer-controlled multiwell-scanning spectrophotometry and robotic and microtray technology. Since the project soon became group oriented, involving the concerted efforts of different expert teams, the results and information of the experiments had to be rapidly generated and shared with the other members of the group. Therefore, laboratory automation was complemented by an in-house-developed information management system that enabled us to capture the experimental data, to process the information immediately, to generate reports, and to store the information in a data base for later retrieval (Figure 4). Finally, the automation also enabled us to limit the type and duration of exposure of the laboratory staff to

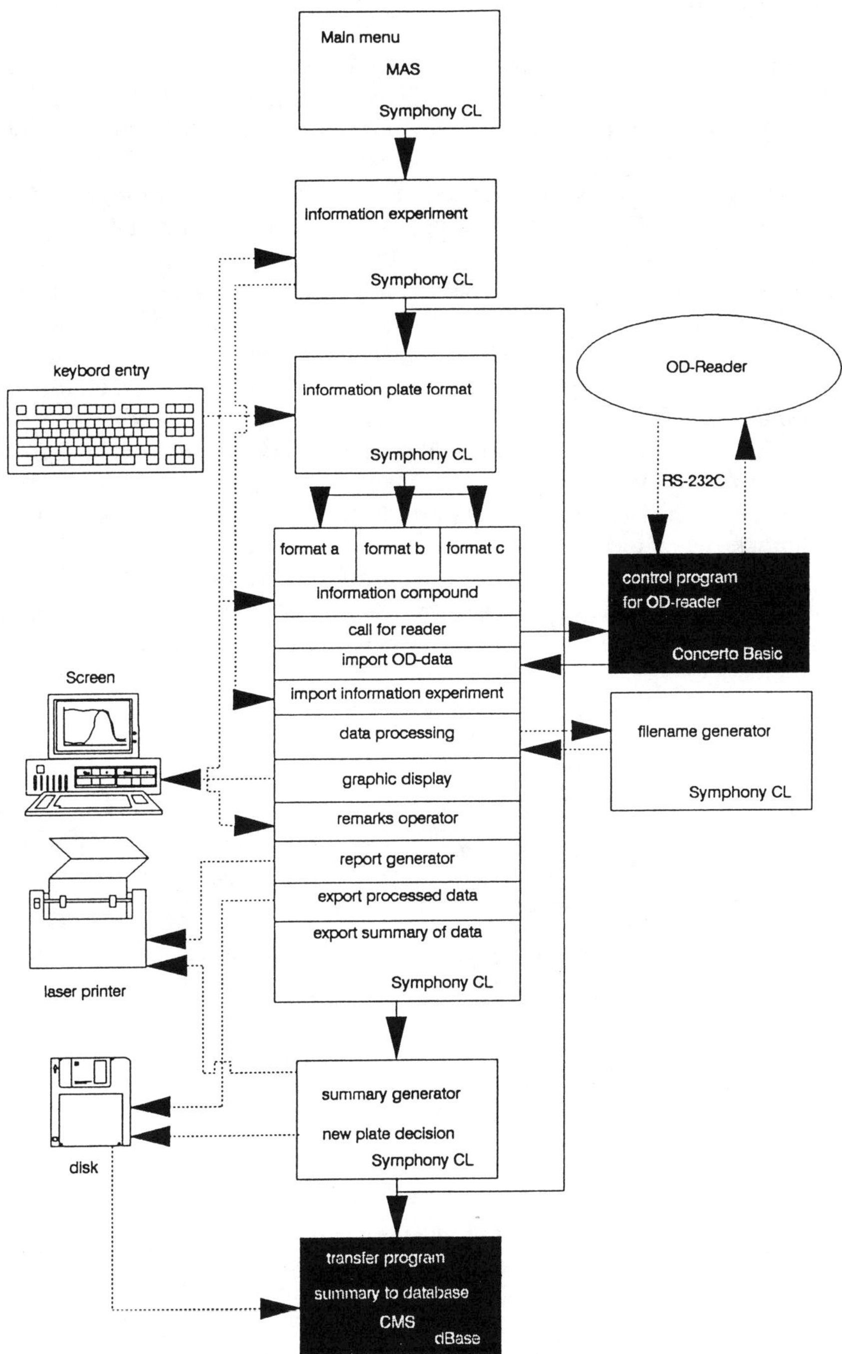

FIGURE 4. Schematic flow diagram of the microtray analyzing system (MAS). Solid lines and arrows represent the flow of the program. The dotted lines indicate the flow of information between the different modules or peripheral devices such as the computers, the laser printer, disks, or the optical density (OD) reader.

potentially biohazardous materials. A general overview of the protocol of the primary evaluation system is given in Table 1. Sterile filling of the microtiter trays with culture medium or cell suspensions (steps 1 and 4) could be most efficiently performed by an eight-channel liquid dispenser. The same tool was also used to dispense the MTT solution in the plates (step 6). Whereas the compounds were added manually to the plates, the fivefold dilutions were performed in the microtiter plate by the robotic Biomek® 1000 Automated Laboratory Workstation (Beckman Instruments), which could be placed inside a biohazard safety cabinet. A number of programs for different dilution protocols were written using the Biomek software. Similarly, a two-step protocol was developed for automatic removal of part of the culture medium and addition of a solution to dissolve the formazan crystals that are generated upon reduction of MTT.

In the standard assay, two compounds were tested per plate with HIV and mock-infected controls for each compound. Each NCE was tested at a concentration of 250, 50, 10, and 2 μg/mL. An additional format with nine dilutions per plate was used to cover a wider concentration range if necessary. As many of the test compounds were poorly soluble in

TABLE 1. General overview of the protocol of the primary evaluation system

Step	Time/96-well plate (seconds)	Tool
(1) Filling of plates with culture medium (100 μL/well)	5	Dispenser[a] (Multidrop®)
(2) Addition of compound to plate (25 μL, 10 × stock solution)	30	Manual pipetting
(3) Serial dilution of compound in plate	30	Robot[b] (Biomek®)
(4) Parallel addition to plates of mock- and HIV-infected cells	5	Dispenser (Multidrop®)
(5) Incubation at 37 °C for 5 days	–	–
(6) Addition of MTT (20 μL/well)	2	Dispenser (Multidrop®)
(7) Incubation at 37 °C for 60 min	–	–
(8) Partial removal of culture medium; replacement by solvent (100 μL/well)	30	Robot (Biomek®)
(9) Dissolution of formazan crystals on plate shaker	30	Four-plate shaker[c]
(10) Data entry in computer (plate identification)	15	computer[d]
(11) Reading of optical density (OD)	5	OD reader[e] (Multiskan®)
(12) Data processing, storage, and reporting	60	computer[d]

[a]Titertek Multidrop®, 8-channel liquid dispenser, Flow Laboratories, Irvine, Scotland.
[b]Biomek® 1000 Automated Laboratory Workstation, Beckman Instruments, Inc., Spinco Division, Palo Alto, California.
[c]Titertek®, Flow Laboratories, Irvine, Scotland.
[d]IBM PS/2 Model 80, software.
[e]Eight-channel optical density reader, Multiskan® MCC 340, Flow Laboratories.

water, stock solutions were routinely prepared in DMSO (dimethyl sulfoxide) at a concentration of 20 mg/mL. Although some compounds still precipitated at the highest concentration(s), we decided not to spend additional time in resolving the solubility problem but rather to concentrate our efforts on testing more compounds. For each concentration, the mean, the median, and the standard deviation based on the optical density values of triplicate wells were determined. These values were then translated into percentage cytotoxicity and percentage antiviral protection, from which the IC_{50}, CC_{50}, and selectivity index were calculated. All this information was available in both graphical and tabular formats. A complete report containing experimental information, the compound's identification, and the results of the calculations was generated within minutes after the analysis of the microtiter plate. After human approval of the results of each compound, summaries were faxed to JRF. Currently, data are transmitted via a modem directly to the host computer in Beerse. To date about 25,000 primary assays have been performed during the JRF and other research projects. Each of these assays represented the determination of a full dose-response (at least 4, triplicate concentrations) of the anti-HIV activity and cytotoxicity of the test compound. All these programs were written in our laboratory, and they were modified according to our growing experience. This rationalization of the procedures has meanwhile led to the introduction of bar codes and optical lasers, and now forms the basis for the development of a fully stand-alone, knowledge-based robotic station for this type of laboratory operation. During the whole project, the primary assay was carried out by two people, who were also involved in other laboratory and organizational activities.

Identification of the Lead Compound

The belief that each new compound tested was potentially a new road to anti-HIV chemotherapy was a driving force that kept us looking in the microscope during the first months of our program. Such a motivation was required in light of all the inactive compounds we were testing. This contrasted sharply with the efforts of our institute in the dideoxynucleoside field, where analogues with interesting activities were often identified. Regularly we noticed some degree of anti-HIV activity with the JRF compounds. However, none of them attained 50% protection when the test was repeated. Inactive compounds were tested only once, as the chances to detect activity are greater when a new compound is tested. In September 1987, barely 3 months after the start of the project, we found that a compound with code number R14458 clearly protected the cells against the HIV-induced CPE at a concentration of 10–50 μg/mL. Trypan blue staining of the viable cells (Figure 5) indicated that the 50% effective concentration for inhibition of viral cytopathicity was 62 μM. The CC_{50} was 612 μM, resulting in a selectivity index of 10. Although this was a moderate activity compared to that of some dideoxynucleoside analogues that we were running as controls, we were immediately excited by our observations. As soon as I communicated the results to Dr. Koen Andries, the finding was received with great enthusiasm in Beerse. Dr. Paul Janssen informed us personally that their records indicated that this compound had been synthesized at JRF back in the early 1960s by Drs. A. Raeymaeckers and J. Van Gelder in an attempt to develop novel types of benzodiazepines. Yet the compound was devoid of diazepam-like pharmacological effects in rodents and in fact had not shown any other pharmacological activity in a series of in vitro and in vivo assays. R14458 was a 6-substituted 4,5,6,7-tetrahydro-5-methylimidazo[4,5,1-*jk*][1,4]benzodiazepin-2(1*H*)-one derivative containing a unique tricyclic ring system (Figure 6); this name would be later abbreviated to TIBO (Pauwels et al., 1990a,b). This structure was also unrelated to any other antiviral agent known at the time. The next week we confirmed the data and started evaluating other compounds of the repository that were related to R14458. Although they were inactive, it merely pointed to the unique observation of R14458. Dr. Paul Janssen very rapidly decided to initiate a new chemical synthesis program based on this new lead

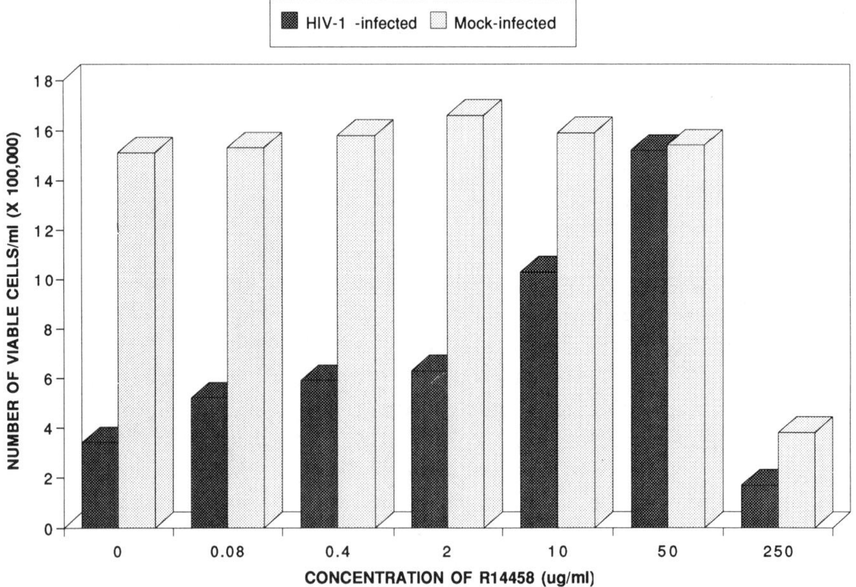

FIGURE 5. Anti-HIV-1 (III$_B$) activity of R14458 in MT-4 cells analyzed by the trypan blue dye exclusion method.

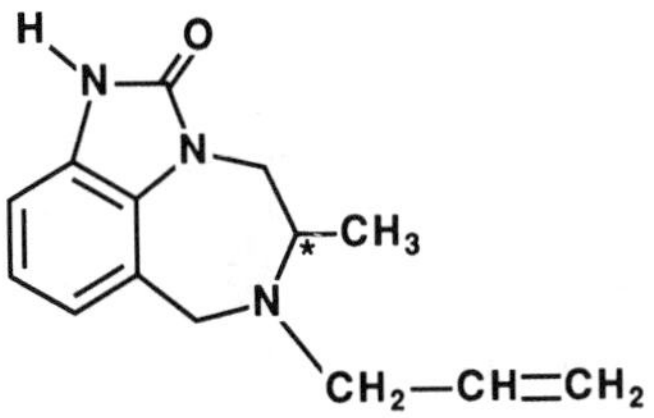

FIGURE 6. Chemical structure of R14458, the first lead compound.

compound. He selected the talented team of Drs. Michael Kukla and Henry Breslin from the JRF in Spring House, Pennsylvania, to synthesize new TIBO analogues. The further screening of compounds of the prototype list would obviously continue. In the meantime our laboratory had become the Centralized Facility of the European Economic Community (EEC) for anti-HIV chemotherapy studies. We were receiving numerous compounds from many companies and institutions in Europe and other parts of the world. At that time our screening capacity was in the range of about 100 microtiter plates per week.

Lead Optimization

The early results with the analogues of R14458 from the repository already pointed to the importance of the substituent in the 6-position. As a result, Drs. Paul Janssen and Marcel Janssen and Dr. Michael Kukla's team decided to synthesize new analogues with different groups appended to the 6-position nitrogen of the diazepin ring (Kukla et al., 1991a). Since R14458 was a racemic mixture, one of the first goals was also to determine whether the anti-HIV activity was enantiospecific at the 5-position. To that end, a common synthetic intermediate was synthesized from which all the TIBO analogues could be derived. The team in Spring House also developed a synthetic enantiospecific chemical pathway allowing the direct synthesis of each enantiomer. In this route, the asymmetric center was introduced via the incorporation of the natural *(S)-* or *(R)-*alanine. The two optical isomers of R14458 were synthesized, R78304 and R78305, with the (−)-*(R)-* and (+)-*(S)-*configuration, respectively. Only the (+)-*(S)-*isomer inhibited HIV-1 replication (Table 2). This requirement for the (+)-*(S)-*configuration for anti-HIV-1 activity was later also observed for all TIBO derivatives that were synthesized. In addition to the fact that some related compounds of R14458 were found inactive, this enantiospecificity lent credence to a specific pharmacological intervention of these compounds at some yet undetermined point in the replicative cycle of HIV. The nearly equal

TABLE 2. Structure–activity relationship for inhibition of HIV-1 cytopathogenicity and cytotoxicity in MT-4 cells by tetrahydro-imidazo[4,5,1-*jk*][1,4]-benzodiazepin-2 (1*H*)-one and -thione (TIBO) derivatives

Compound	Code	R_1	R_2	R_3	R_4	R_5	R_6	IC_{50}^{a} (nM)	CC_{50}^{b} (μM)	SI^{c}	Relative potency
R14458	a	$-CH_2-CH=CH_2$	$(\pm)$-CH_3	H	O	H	H	59,348	621	10	1
R78305	b	$-CH_2-CH=CH_2$	$(+)$-(S)-CH_3	H	O	H	H	66,880	681	10	1
R78819	c	$-CH_2-\overset{CH_3}{C}=CH_2$	$(+)-(S)-CH_3$	H	O	H	H	13,676	112	8	4
R80902	d	$-CH_2-CH=C\overset{CH_3}{\underset{CH_3}{}}$	$(+)-(S)-CH_3$	H	O	H	H	4,207	470	112	14
R81886	e	$-CH_2-\overset{CH_3}{C}=CH_2$	$(+)-CH_3$	H	O	9-Cl	H	4,702	17	4	13
R86080	f	$-CH_2-CH=C\overset{CH_2CH_3}{\underset{CH_2CH_3}{}}$	$(\pm)-CH_3$	H	O	9-CH_3	H	314	20	64	189
R85386	g	$-CH_2-CH=C\overset{CH_2CH_3}{\underset{CH_2CH_3}{}}$	$(-)-(S)-CH_3$	H	O	9-Cl	H	25	22	880	2,374

R82150	*h*	$-CH_2-CH=C(CH_3)_2$	$(+)-(S)-CH_3$	H	S	H	H	44	552	12,545	1,349
R84963	*i*	$-CH_2-CH=C(CH_3)_2$	$(\pm)-CH_3$	H	S	H	$CH_3(trans)$	39	81	2,077	1,522
R84914	*j*	$-CH_2-CH=C(CH_3)_2$	$(\pm)-CH_3$	H	S	H	$CH_3\ (cis)$	787	74	94	75
R80806	*k*	$-CH_2-CH_2-CH_3$	$(\pm)-CH_3$	H	S	H	H	242	556	2,339	245
R82913	*l*	$-CH_2-CH=C(CH_3)_2$	$(+)-(S)-CH_3$	H	S	9-Cl	H	33	34	1,030	1,798
R86085	*m*	$-CH_2-CH_2CH(\text{cyclopropyl})$	$(-)-(S)-CH_3$	H	S	9-Cl	H	334	36	108	178
R86777	*n*	$-CH_2-CH_2-CH(CH_3)_2$	$(-)-(S)-CH_3$	H	S	9-Cl	H	92	19	207	645
R85787	*o*	$-CH_2-CH(\text{cyclobutyl})$	$(-)-(S)-CH_3$	H	S	9-Cl	H	43	28	651	1,380
R86162	*p*	$-CH_2-CH=C(CH_2CH_3)_2$	$(-)-(S)-CH_3$	H	S	9-Cl	H	15	12	800	3,957
R86154	*q*	$-CH_2-C(CH_3)=CH_2$	H	$(+)-S-CH_3$	S	9-Cl	H	446	24	54	133
R86150	*r*	$-CH_2-CH(\text{cyclopropyl})$	H	$(-)-(R)-CH_3$	S	9-Cl	H	2,216	>812	>366	27

(*continued*)

TABLE 2. (*continued*)

Compound	Code	R_1	R_2	R_3	R_4	R_5	R_6	IC_{50}[a] (nM)	CC_{50}[b] (μM)	SI[c]	Relative potency
R85255	s	$-CH_2-CH=C(CH_3)CH_3$	$(+)-(S)-CH_3$	H	S	9-Cl,10-Cl	H	25	45	1,800	2,374
R86183	t	$-CH_2-CH=C(CH_3)CH_3$	$(+)-(S)-CH_3$	H	S	8-Cl	H	4.6	138	30,000	12,902
R87027	u	$-CH_2-CH=C(CH_2CH_3)CH_2CH_3$	$(+)-(S)-CH_3$	H	S	8-Cl	H	5.1	12	2,353	11,637
R86775	v	$-CH_2-CH=C(CH_3)CH_3$	$(+)-(S)-CH_3$	H	S	8-Br	H	3.0	53	17,667	19,783
R84674	w	$-CH_2-CH=C(CH_3)CH_3$	$(+)-(S)-CH_3$	H	S	8-CH_3	H	14	80	5,714	4,239

[a]Fifty percent antiviral inhibitory concentration, or concentration required to achieve 50% protection of the MT-4 cells against HIV-1-induced cytopathicity as determined 5 days postinfection by the MTT procedure.
[b]Fifty percent cytotoxic concentration, or concentration required to reduce the viability of mock-infected MT-4 cells by 50% after 5 days incubation in the presence of the compound.
[c]Selectivity index, or ratio of CC_{50} to IC_{50}.

cytotoxicity of the two compounds provided further evidence that the anti-HIV activity and cytotoxicity were not linked and could vary independently. After about 50 new TIBO analogues were synthesized, the optimal features were combined in the dimethylallyl analogue of R14458, the (+)-(S)-isomer R80902. Both the potency and the selectivity of this compound were about 10-fold improved over the initially discovered lead compound. However, more pronounced improvements were obtained during the structure–activity relationship (SAR) study of analogues with variations of the five-membered urea ring. This period, characterized by the evolution from moderately active to very potent inhibitors of HIV replication, was certainly one of the most exciting episodes of the project. Our primary screening system had by that time reached its cruising speed, so that the flux of information from the antiviral unit to the chemists was rapid. Compounds were often shipped by Dr. Kukla's team on Tuesdays, arriving in Belgium the next morning. A company car would then take the compounds to JRF in Beerse, where on Wednesday and Thursday the chemical identity and purity were confirmed, an essential step in the life cycle of the Janssen compounds in order to obtain the "R-number" registration. On Friday mornings, the compounds arrived in the HIV laboratory in Leuven, where a test was immediately initiated. Five days later the anti-HIV data were communicated to Dr. Koen Andries. It was not exceptional that Dr. Paul Janssen, who was a daily visitor of Dr. Andries, called us a few hours later and inquired about the details of the experiment. Having discussed the new results with the head of the chemistry department, Dr. Marcel Janssen, Dr. Paul Janssen often called Dr. Kukla that same day suggesting new directions in the program. From then on my daily telephone conversations with Dr. Andries became legendary in the laboratory.

Although many different rings were synthesized during this period and found to be inactive, replacement of the urea oxygen by sulfur at the C-2 position of the imidazolone ring resulted in compounds which were several 100-fold more active than the original lead compound (Kukla et al., 1991b). The TIBO derivative R82150 [or (+)-(S)-4,5,6,7-tetrahydro-5-methyl-6-(3-methyl-2-butenyl)imidazo[4,5,1-jk][1,4]-benzodiazepin-2 (1H)-thione] (Figure 7) inhibited HIV-1 replication at a concentration of about 30 nM (Figure 8). This high level of potency was not paralleled with increased toxicity to the host cells. As a consequence, the compound routinely achieved selectivity indexes from 10,000 to 50,000. At that time, none of the other HIV-1 inhibitors we were working with in our institute or which had been published elsewhere had ever attained a selectivity index of this magnitude. In 1989, we decided to publish our findings (Pauwels et al., 1990a,b). During the preparation of the paper (1990a), another molecule, R82913, was synthesized with a chlorine at the C-9 position of the phenyl moiety. This TIBO derivative was repeatedly active in the nanomolar range with IC_{50} values around 1–5

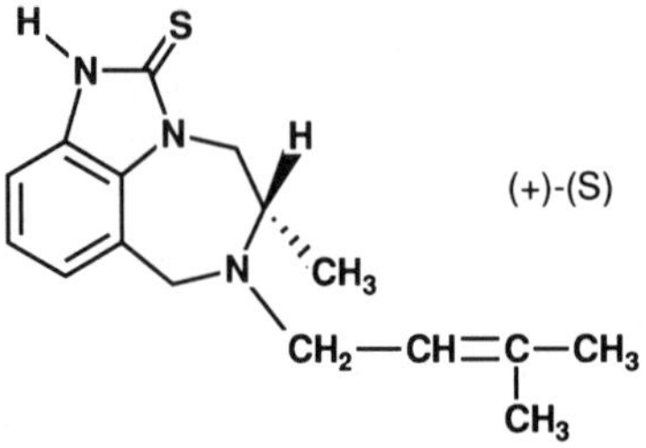

$(+)\text{-}(S)$

$CH_2-CH{=}C-CH_3$
$|$
CH_3

FIGURE 7. Chemical structure of TIBO derivative R82150.

nM. Several months later, by changing the assay conditions of our primary screen, we observed a striking variation in IC_{50} values that seemed to depend on the starting concentration used in the microtiter plates. This came as a surprise since neither one of our traditional controls, such as AZT, DDC, DDI, or dextran sulfate, nor our "prototype" TIBO derivative R82150, had exhibited such a behavior. Our original standard dilution protocol, set up with the Biomek® robot station, encompassed the use of a single set of 200 μL tips to prepare up to nine 5-fold dilutions in 96-well plates using 25-μL transfer volumes in 100-μL dilution medium and 75-μL mixing volumes. This protocol had been validated with a radiolabeled compound. However, HPLC (high-pressure liquid chromatography) analysis of drug samples diluted in 96-well trays now revealed that R82913 exhibited a strong plastic adherent effect. If a single tip was used, the compound which had adhered to the plastic leaked back to the wells at higher dilutions, leading to an underestimation of the actual drug concentration and thus an overestimation of the activity. Our studies also indicated that when tips were changed after each dilution, the opposite occurred, i.e., compound was lost through adherence to the disposed plastic tips, thus leading to overestimation of the actual drug concentration and underestimation of the activity. The phenomenon was restricted to highly hydrophobic substances with high activity (below 100 nM). Owing to the safety precautions in our facility, glass containers and needles could not be used to circumvent this problem. Therefore, we determined experimentally that by using a dilution protocol in which tips were changed every three dilution steps, up to 100,000-fold dilutions could be prepared that accurately reflected the theoretical concentrations of both hydrophylic and hydrophobic compounds. From then on all our screening would be performed using this new dilution protocol. On some occasions, 96-well plates were prepared twice so that the concentrations of the wells containing an important analogue could be double-checked by HPLC. So far, the results with the new dilution protocol have been very satisfying. Using this method we have also been able to determine the actual IC_{50} value of R82913, which was about 20–30 nM, a potency similar to that of R82150.

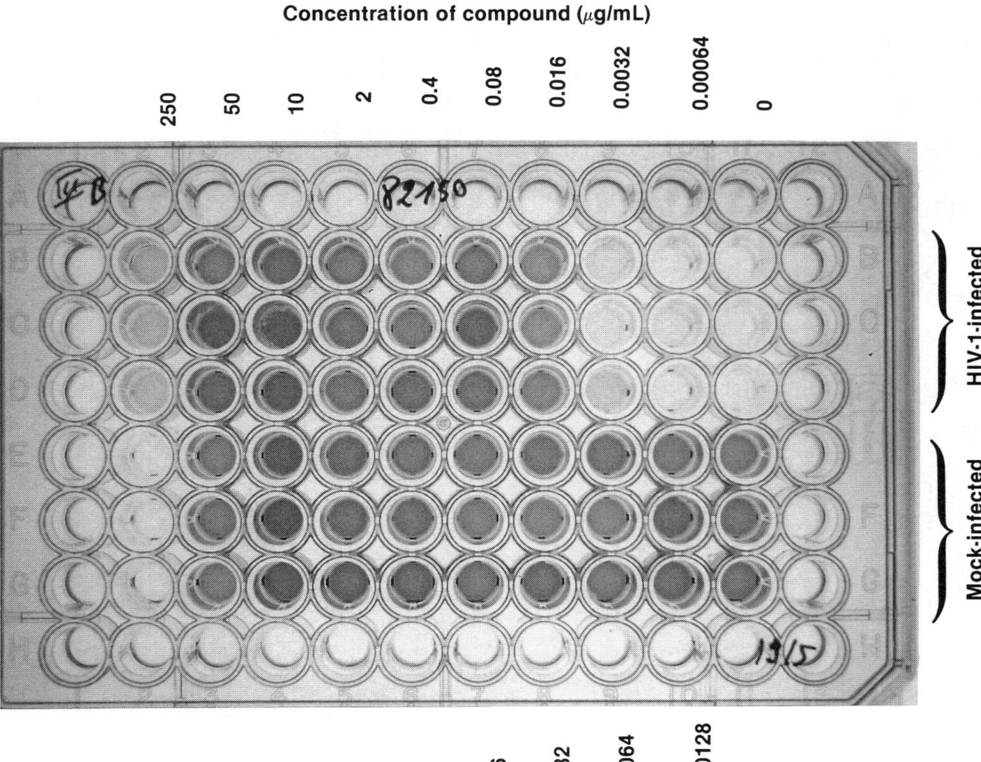

FIGURE 8. Anti-HIV-1 (III$_B$) activity of R82150 in MT-4 cells analyzed by the MTT procedure.

More recently, a new series of TIBO derivative were synthesized with a modification in the 8-position of the phenyl moiety. The 8-chloro derivative, R86183, emerged as one of the most potent and selective compounds, with IC_{50}'s of 1–5 nM and a selectivity index of about 10,000 to 100,000. This meant that the potency of the original lead compound R14458 had been optimized by a factor of over 10,000 with a neglectible change in cytotoxicity. The latest SAR studies also revealed some new features of the TIBO pharmacophore, i.e., the stereochemical requirement of a C-7-methyl substituent (Table 2).

Despite the fact that the majority of the TIBO compounds required a 10–15 step synthesis, this successful lead optimization phase, involving several hundred analogues, was accomplished in about two years following the original discovery of the lead compound.

Antiviral Profile of TIBO Compounds

As the potency of the various TIBO compounds increased during the lead optimization phase, we regularly determined the activity of selected compounds in other antiviral assay systems. As soon as R82150 was discovered, it became a prime candidate for further development. Therefore, in parallel with the first in vivo toxicity studies, the activity of R82150 was further evaluated for its inhibitory effects on various measures of HIV-1 replication in MT-4 cells. A dose-dependent inhibition was found when we monitored production of infectious progeny virus and p24 core protein, plaque formation, and the expression of viral antigens (Figure 9). R82150 was also examined for its activity against HIV-1 (III_B strain) in the non-HTLV-I-carrying CD4 + cell lines CEM and MOLT-4 (Clone 8), in phytohemagglutinin (PHA)-stimulated peripheral blood lymphocytes and in macrophages. Despite the fact that different end points and incubation times were used in these assays, it was clear that the anti-HIV-1 activity was host cell independent. In contrast, the 2′,3′-dideoxynucleoside analogues AZT and ddC (2′,3′-dideoxycitidine) had shown some variations in anti-HIV-1 activity and cytotoxicity depending on the choice of the cell type.

In 1986, the Pasteur group had identified a different type of HIV, HIV-2, in AIDS patients of West African origin (Clavel et al., 1986). As a result of a suggestion of Prof. Jan Desmyter that we needed to expand our collection of HIV strains, I went to Paris and obtained a sample of HIV-2. As soon as we made the necessary virus stocks and had set up an assay in MT-4 cells using the MTT procedure, we determined the anti-HIV-2 activity of representatives of the different classes of compounds that were at the time studied in our laboratory. To our surprise, there were two compounds, one of which was a TIBO compound, that

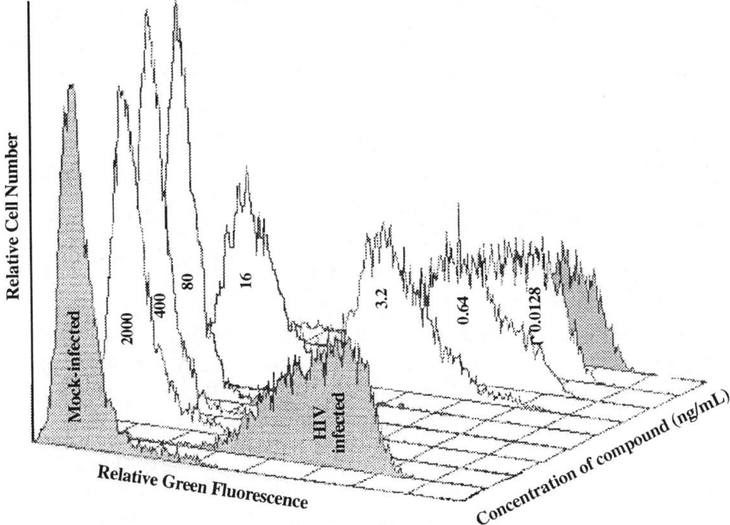

FIGURE 9. Effect of varying concentrations of R82150 on HIV-1 antigen expression in MT-4 cells at 4 days after infection, as determined by laser flow cytofluorometry (FACSTAR, Becton Dickinson, Mountain View, CA). HIV-1 antigens were revealed by an indirect immunofluorescence staining procedure, using polyclonal antibodies as probe. Histograms represent data of a typical experiment based on the analysis of 10,000 cells per sample.

were not active against HIV-2. This was in sharp contrast with the 2′,3′-dideoxynucleoside analogues AZT and ddC, the phosphonate derivative PMEA, and several sulfated polysaccharides, all of which exhibited activities similar to those previously found against HIV-1. The other inactive compound was the 6-substituted acyclouridine derivative, HEPT (Figure 10). This acyclic nucleoside analogue had been synthesized by Dr. H. Tanaka, who worked in the laboratory of Prof. T. Miyasaka (Showa University, Tokyo, Japan) and later in the laboratory of Dr. D. Walker (Birmingham, UK). The HEPT derivative was among

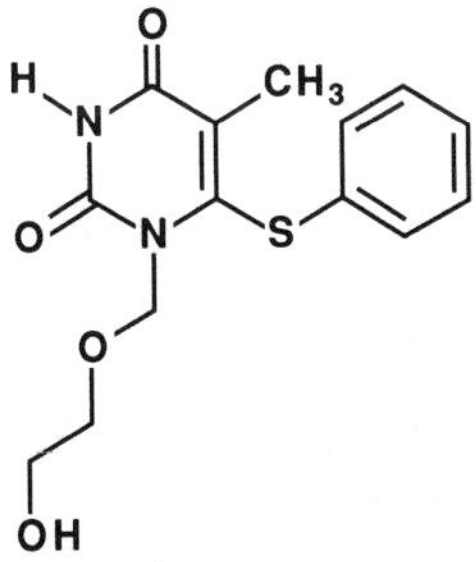

FIGURE 10. HEPT, 1-[(2-dydroxyethoxy)methyl]-6-(phenylthio)thymine.

a series of compounds which had been sent to our institute for anti-herpes evaluation. They were found inactive against herpesviruses, but—bearing in mind that testing compounds against a wide variety of viruses had in the past led to other interesting discoveries—Prof. E. De Clercq asked Dr. M. Baba and myself to determine also the anti-HIV activity in the primary screening assay. We found a moderate anti-HIV activity (IC_{50}: 5–10 μM), with a selectivity index of about 100 (Baba et al., 1989). This meant that by coincidence the HEPT compound was discovered only a few weeks after identification of the original TIBO lead. The mechanism of action of this "nucleoside analogue" was still obscure at the time of discovery of the HIV-1 specificity. A puzzling observation was the fact that the triphosphate species was inactive as an inhibitor of the reverse transcriptase. Now, the apparent HIV-1 specificity brought both compounds under the same spotlight. Some similarities in both the TIBO and HEPT structure were suggested at the time, but there was no general consensus on this issue. Only later, when the mechanism of action of the TIBO compounds was unraveled, did it became clear and accepted that both compounds had indeed a common target. The HIV-1 specificity prompted us to investigate a wide range of retroviruses. All the HIV-1 strains tested, including clinical isolates and laboratory strains originating from different regions of the world, were all susceptible to TIBO inhibition. Yet, none of the Simian (SIV-2, SIV), feline (FIV), or murine retroviruses tested was inhibited by these compounds. Only very recently, we discovered an SIV strain, isolated from African green monkeys, that proved clearly susceptible to TIBO, albeit to a lesser extent than HIV-1 (Debyser et al., 1992b; Pauwels et al., 1992). This observation indicated that the TIBO derivatives were not only highly selective inhibitors of HIV-1 but that they were also endowed with an unprecedented antiretroviral specificity. This hallmark feature led us to speculate that the compounds interacted in a novel way with the replicative cycle of HIV-1. The differential sensitivity of HIV-1 and HIV-2 to the TIBO derivatives would be of crucial importance in the elucidation of the mechanism of action of these compounds.

Mechanism of Action of TIBO

As already outlined in the discussion of our screening strategy, studies of the mechanism of action were only initiated at a later stage of the project. As we started from scratch it seemed more important to us to concentrate our efforts on finding interesting anti-HIV activity first. But once this activity was found and then optimized we became more and more intrigued about the potential site of interaction of this class of compounds. The stringent SAR, the stereospecificity, and later the

HIV-1 specificity, all suggested a defined interaction with a putative receptor structure.

Measuring reverse transcriptase activity was one of the first procedures that we had set up in our laboratory to monitor HIV replication in different cells and to demonstrate RT inhibition by various inhibitors such as suramin and 2′,3′-dideoxynucleosides. Therefore, early on in the project, the activity of R14458 was evaluated in an RT reaction with poly rA • oligo dT as the template/primer. The inhibition observed at about 100 μg/mL was 10-fold higher than the IC_{50} in cell culture and considered to be a borderline activity. Later, as a result of a discussion with Dr. Koen Andries, we tried to set up so-called time-of-addition (TOA) experiments to delineate the drug-sensitive phase of the TIBO compounds. In the TOA experiment (Figure 11) MT-4 cells were infected with HIV-1 at a high multiplicity of infection (MOI), so that viruses

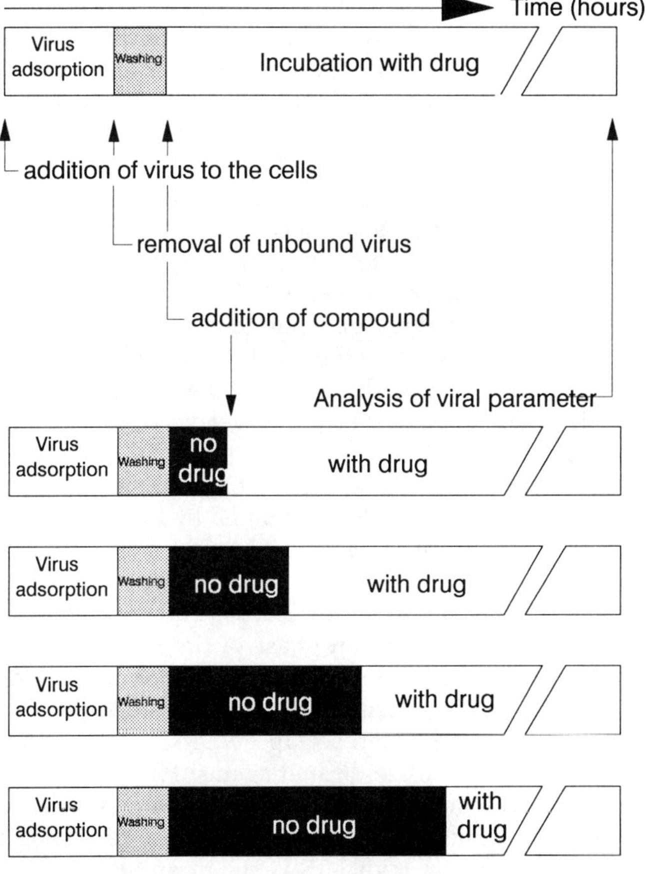

FIGURE 11. Scheme of time-of-addition (TOA) experiment.

advance, in principle synchronously, through the different stages of the replicative cycle. When a protective compound is added at the time of infection, no viral replication can take place. However, if addition of the compound is systematically delayed, protection is observed up to the moment that the virus has passed the stage at which the antiviral compound interacts. Obviously, the resolution of this assay can be dramatically increased if reference compounds with known modes of action are included. The TOA experiment is therefore an ideal first-line experiment, as it concentrates on identifying the target that is actually responsible for the inhibition of viral replication in cell culture, rather than whether any particular step of the cycle is inhibited or not. However, as large quantities of HIV were needed and compounds were used at equipotent concentrations, the assay works best if compounds have a selectivity index of at least 50–100. In the first experiments pentosan polysulfate, aurintricarboxylic acid, suramin, AZT, ddC, and ddI (2′,3′-dideoxyinosine) were run in parallel. Pentosan polysulfate, which was known to interfere with virus binding, was no longer able to inhibit HIV-1 replication if added 1.5 hours (or later) after infection (Figure 12). Also aurintricarboxylic acid and suramin completely lost their protective effect if added directly after infection (or later). This suggests that, despite the fact these agents had been shown previously to inhibit RT, their actual mode of action would be intervention with a very early step, possibly virus binding. In contrast, ddC, AZT, and ddI only began to lose their inhibitory effect in a window that extended from 3 to 6 hours after infection. This variation may be related to differences in the time that is needed to activate the 2′,3′-dideoxynucleoside (ddN) analogues during the various phosporylation steps. The fact that the TOA curves for R82913 and HEPT coincided with that of ddI was suggestive of interaction with an event also affected by the ddN analogues. The idea of RT being the target was not immediately and generally accepted. Additional experiments were carried out to substantiate the RT hypothesis. Both the TIBO and the HEPT derivative inhibited newly formed HIV DNA amplified by PCR (polymerase chain reaction) and isolated from MT-4 cells infected under similar conditions as for the TOA experiment. This observation, as well as the absence of activity in persistently HIV-infected T-cells, indicated that they indeed interacted with the preintegration phase of the replicative cycle. Other options such as potential effects on virus binding, masking of the CD4 receptor, and syncytium formation were excluded via a series of functional assays that had been set up by Dr. D. Schols. As only one TIBO compound had ever been tested against HIV-1 RT, I asked Zeger Debyser, a doctoral student who had gained some experience with RT in our laboratory during his graduate studies, to carry out a number of additional experiments with RT. First, an anti-RT experiment was carried out with the more potent TIBO derivative R82150. Similarly to

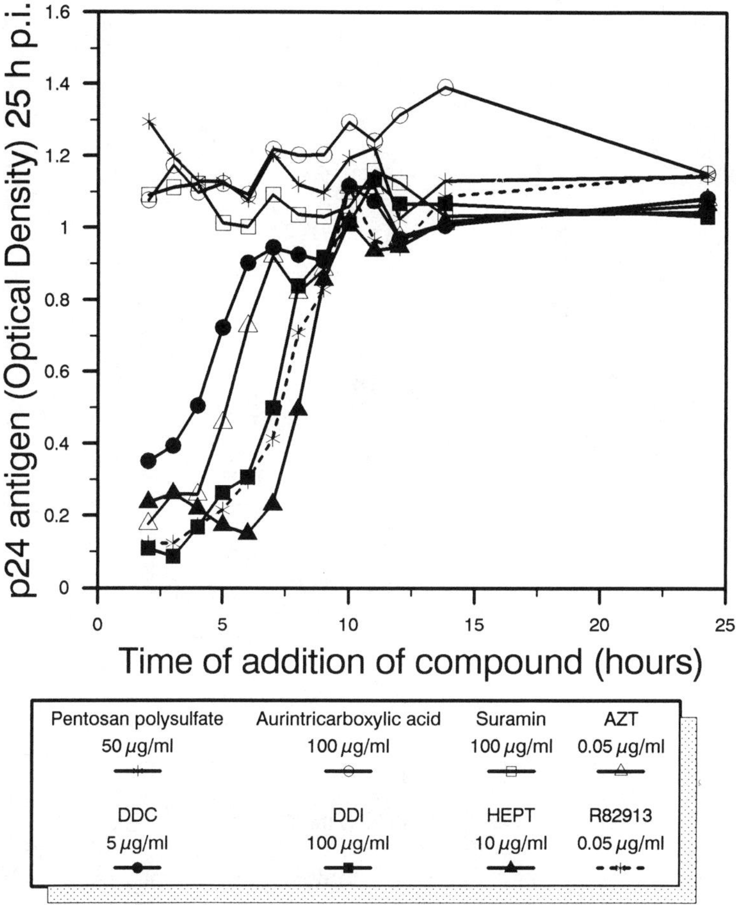

FIGURE 12. TOA experiment. MT-4 cells were infected at a high multiplicity of infection with HIV-1 (HTLV III$_B$). After 60-min incubation at 37 °C unabsorbed virus was removed by three washing steps. HIV-1 core protein (p24) was determined 25 hours postinfection by an antigen-capture assay using a sandwich ELISA (enzyme-linked immunosorbent assay) technique.

R14458, the IC$_{50}$ for RT inhibition with poly rA • oligo dT as the template/primer was significantly higher than that for inhibition in cell culture. Yet, the absolute value (about 5 μM) was too low to be coincidental. It was subsequently demonstrated that, when the inhibitory effects of a series of TIBO compounds on HIV-1 replication in cell culture and their effects on HIV-1 recombinant p66 RT were compared, a close correlation emerged (Figure 13). Also the same stereospecificity was observed at the RT level. In contrast, the IC$_{50}$ values for HIV-1 replication did not correlate with the cytotoxic effects (CC$_{50}$ values) of the TIBO compounds. In agreement with the HIV-1 specificity in cell

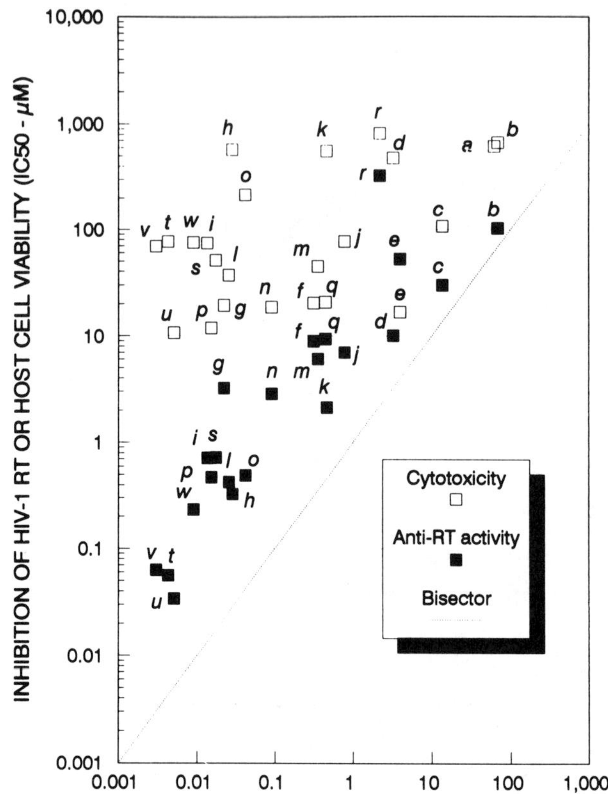

FIGURE 13. Correlation between HIV-1 inhibitory effects of a series of TIBO derivatives in cell culture and inhibition of HIV-1 recombinant reverse transcriptase (RT); for an explanation of codes *a–w*, see Table 2.

culture, we also found no inhibitory effects of R82150 at concentrations up to 350 μM on RTs derived from HIV-2, AMV (avian myeloblastosis virus), and MLV (Moloney murine leukemia virus). All these experiments clearly demonstrated that RT was indeed the target of this class of compounds. From then on, Zeger Debyser would also become fully engaged in the study of the mechanism of RT inhibition by TIBO compounds.

We were gratified to learn that experiments by Dr. K. Ono (Nagoya, Japan) indicated that the human DNA polymerases α, β, or γ were totally refractory to TIBO inhibition. This was in sharp contrast with many other RT inhibitors at the time, which despite some degree of selectivity always had some residual activity on these enzymes. Detailed analysis of RT inhibition subsequently showed that RT inhibition was increased by about 15-fold when poly rC • oligo dG_{12-18} was used as the template/primer. The template preference would be a typical feature of this class of compounds, which also led to selective inhibition of the RNA-dependent DNA polymerase catalytic function of RT. Kinetic studies with the template/primer revealed an uncompetitive type of inhibition, suggesting an ordered mechanism of inhibition whereby binding of the inhibitor is preceded by binding of the template/primer to the enzyme (Debyser et al., 1991). Owing to HIV-1 specificity and the common behavior in the TOA experiment, many of these experiments were also carried out with the HEPT compounds, which clearly indicated that they had a common target. Whereas the inhibition by TIBO was found to be noncompetitive with respect to the different natural substrates, inhibition of RT by some HEPT congeners was noncompetitive with respect to dGTP but competitive with respect to dTTP (Baba et al., 1991). Having reviewed all our data and in an attempt to explain all our observations, we speculated that the allosteric TIBO-binding site would be a pocket, essentially hydrophobic in nature, located in the immediate vicinity of the substrate-binding site (Debyser et al., 1991), a model that was also presented at the Fourth International Conference on Antiviral Research in New Orleans in April 1991.

As these experiments proceeded several other groups reported on compounds with a remarkable "TIBO-like" antiviral profile. These included the dipyridodiazepinone (Merluzzi et al., 1990; see also Chapter 3 by Adams and Merluzzi), pyridinone (Goldman et al., 1991; see also Chapter 5 by Goldman) and bis(heteroaryl)piperazine (BHAP) derivatives (Romero et al., 1991). All three groups of compounds were identified through empirical screening in RT assays. In 1990, we discovered a new lead compound, an α-anilinophenylacetamide (α-APA) (Figure 14) derivative that was rapidly optimized to potencies and selectivities comparable to that of TIBO (Pauwels et al., 1993). Also, these compounds, including α-APA, shared the unique HIV-1 specificity, similar template/primer preferences, and were, with the exception of some HEPT derivatives, noncompetitive inhibitors with respect to the substrate. Further evidence that these compounds had a common mechanism of action stemmed from the fact that TIBO compounds were

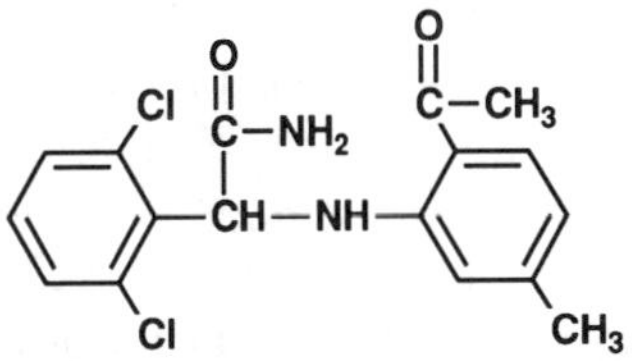

R 89439

FIGURE 14. Chemical structure of α-APA derivative R89439.

able to protect HIV-1 RT from photoinactivation by an azidodipyrido-diazepinone analogue (Wu et al., 1991) and that HIV-1 mutants that are 1000-fold less sensitive to pyridinone derivatives are cross-resistant to TIBO and dipyridodiazepinone (Nunberg et al., 1991). Sequence analysis of these HIV-1 mutants and photoadducts as well the preparation of chimeric RT constructs (Shih et al., 1991), as well as the observations of site-directed mutagenesis studies by Karen De Vreese in our laboratory (De Vreese et al., 1992), all pointed to the importance of the tyrosine residues at amino acid positions 181 and 188 for interaction with TIBO and TIBO-like compounds. Current work in our laboratory by Karen De Vreese and Dr. Anne-Mieke Vandamme is concentrated on the further characterization of viral strains resistant to different types of TIBO-like compounds in order to get better insight into the role of the different amino acids that constitute the "TIBO pocket." An ongoing collaboration with the team of Dr. E. Arnold at Rutgers University (New Jersey), where cocrystallization studies are being performed may lead to the visualization of this site, which certainly will provide new insights into the molecular mechanisms by which these compounds inhibit the reverse transcription process.

Development of a Clinical Candidate

The criteria that governed the selection process for a clinical candidate were in principle very simple. What we needed was a nontoxic compound that, upon administration, achieved sustained plasma levels superior to the in vitro IC_{90} values or roughly 5 IC_{50} equivalents. The idea started to materialize as soon as the potent anti-HIV activity of R82150 was demonstrated in cell culture in May 1989. We immediately initiated additional studies in other anti-HIV systems and investigated the cytotoxic properties in various T-cell models, monocyte/macrophages, fibroblast cell lines, and bone-marrow progenitor cells. In Beerse, the compound was rapidly submitted to a whole panel of different pharmacological assays. They all confirmed the extremely low toxicity profile of TIBO, and the compound was classified as pharmacologically "inert." This was of course not totally unexpected because of the criteria used to select the original prototype compounds. The JRF team of Drs. Jos Heykants and Robert Woestenborghs started studying the pharmacokinetics of R82150 in beagle. They found that shortly after a 1 mg/kg bolus injection, plasma levels of R82150 exceeding 1 μg/mL were obtained. The terminal half-life was 3.2 hours, and the compound was found to be extensively distributed to the tissues. Cardiovascular and behavioral studies, mainly in Labrador retrievers, did not reveal any side effects.

The two first volunteers to take R82150 were Dr. Geert Cauwenbergh, one of the main clinical investigators at JRF and coordinator of the planned TIBO clinical trials, and Dr. Koen Andries. The pharmacokinetics and absolute bioavailability of R82150 were further investigated in six healthy male volunteers. The fact that the different metabolic patterns of several of these individuals was known from previous studies helped to place the pharmacokinetic findings in a wider context. After oral administration of 100- and 200-mg doses, the absorption of the test compound was rapid, since peak plasma levels were attained within 0.5–1 hour in all subjects. Peak plasma levels averaged 0.14 and 0.38 mg/mL, respectively. Plasma levels of R82150, exceeding the antiviral IC_{50}, could clearly be maintained for at least 8 hours with the 100-mg oral dose. Terminal half-lives after oral administration were on the order of 11–15 hours. The dose-normalized absolute bioavailability of R82150 was 24% for the 100-mg dose and 31% for the 200-mg dose. Clinically relevant changes in hematological, biochemical, and cardiovascular variables were not seen in any of the subjects at any of the study phases, and no subjective side effects were reported. The extensive tissue distribution, the relatively long half-life, and the acceptable oral bioavailability were features of a favorable pharmacokinetic profile of the prototype compound R82150.

In August 1989, R82913 was synthesized and found to be apparently even more active than R82150. At one of the informal meetings of the TIBO team in his house in Beerse, Dr. Paul Janssen suggested that we needed to know as soon as possible whether R82913 was a type of compound that could be given to patients without the occurrence of side effects over longer periods of time and whether plasma levels could be sustained that exceeded the in vitro protective concentrations. It was clear of course that, owing to the HIV-1 specificity, the TIBO compounds could not be tested in animal antiretroviral models. Several teams at JRF subsequently worked day in and day out studying the animal toxicicity profiles, metabolism, and pharmacokinetics of R82913 by the route that had been followed for R82150. A few months later, all the necessary data for a Phase I trial were presented to Dr. Brian Gazzard of the Kobler Institute in London, who had been invited to Beerse by Dr. Geert Cauwenbergh. Dr. Gazzard agreed to present the data to the ethical committee of his institute, which soon approved a small trial in AIDS patients.

A major handicap for the development of the TIBO compounds was the complex synthesis, which also involved a dangerous (on large scale) reduction reaction with lithium aluminum hydride. Despite these difficulties, the team of Dr. Jef Dockx, who was responsible for scaling up the TIBO synthesis, managed to prepare the quantities necessary for the trial. Around Christmas 1989, barely 6 months after its initial synthesis, R82913 was administered to the first AIDS patients. Some weeks later, a similar trial would start in Paris under the supervision of Dr. Gilles Pialoux. In both studies, the intravenous route was choosen because R82913 had a somewhat lower oral bioavailability than R82150, only limited amount of compound was available, and a better control of the compliance was possible. In January 1990, Dr. Paul Janssen invited several of us to join him on a visit to London, where we had the unique opportunity to talk with the patients who were receiving the TIBO compound. Although the primary goal of the study was to assess pharmacokinetics and potential side effects, the study protocol had also included the measurement of HIV p24 core protein and CD4 counts, two parameters for which pretherapy values were available. Although our so-called miracle scenario did not happen, the overall change in p24, which seemed to correlate with higher trough levels, suggested some anti-HIV activity in vivo (Pialoux et al., 1991). The study also demonstrated that TIBO compounds were indeed very well tolerated, even after prolonged treatment. Since this study new TIBO derivatives have been found with improved anti-HIV activity. One of these compounds, R86183, is now considered as a new potential clinical candidate. However, since the large-scale synthesis of this compound is still more complex than that of the newly discovered α-APA derivatives, the clinical studies are currently more focused on the latter group of compounds.

Conclusion and Future Prospects

Our ultimate goal, the development of a clinically effective anti-HIV drug has not yet been fully achieved. However, the discovery of the TIBO derivatives as the first prototype of a new generation of ultraspecific antivirals offers new possibilities for the development of effective anti-HIV agents. Under in vivo conditions, this high specificity is translated into low or no toxicity, a factor which had in the past often impeded the clinical development of earlier generations of "broad-spectrum" antiviral agents. Several clinical trials with TIBO-like compounds are currently underway and should demonstrate whether this class of compounds can or cannot benefit patients with HIV infections. To date some preliminary findings of these studies seem to suggest that HIV replication can efficiently be stopped in the short term but that resistent HIV-1 strains rapidly emerge. It remains to be determined whether these virus variants are as pathogenic as the original pool of susceptible strains or not. If indeed the resistance would be a limitation for their use as single agents, the possibilities of some form of combination with another type of anti-HIV agent should be fully investigated. It is also important to note that recent findings in our laboratory indicate that resistance to a particular TIBO-like compound does not automatically lead to cross-resistance to other nonnucleoside RT inhibitors. These observations therefore suggest that inside the binding site for TIBO-like compounds (the "TIBO pocket"), these compounds, depending on their chemical structure, display quantitative and/or qualitative differences in their interactions with the amino acids that constitute this putative pocket, and may induce different conformational changes. As we learn more about this "pocket" on RT, it is not inconceivable that, by using novel screening strategies and new insights gained from crystallographic and mechanistic studies, a second generation of RT inhibitors might be developed that would have a somewhat broader specificity and hence would also cover the spectrum of HIV strains which now escape from the current drugs. After all, what we, with cell culture assays, have been doing is to define our selection criteria on one strain only and as a consequence we have identified and optimized toward that particular strain. The real impact of HIV being a quasi-species was only recognized by us later. A number of observations in this area are also reminescent of findings in the antirhinovirus field (see Chapter 8 of this volume by Andries), where the serotype specificity of antiviral agents, the rapid resistance to narrow-spectrum agents, and also a (hydrophobic) pocket have been recognized for some time.

Although we have identified in our cellular assays some compounds directed at other targets, it remains puzzling why many groups are finding compounds of a different chemical nature, directed at the same target and even at the same site on that target. Apparently this points to a highly vulnerable spot of the HIV replicative cycle that is located in the RT at a location in the immediate vicinity of the catalytic center. It will be interesting to learn what role this natural "chemotherapeutic" niche plays in the catalytic events leading to reverse transcription and whether similar processes occur in other polymerases of cellular, bacterial, and viral origin.

Particularly during the early phase of our project, I and other members of the TIBO team were sometimes questioned as to whether "just" screening compounds "from the shelf" was indeed the right approach and whether it was not merely waiting for a lucky hit. Also, as the virology laboratory was located in an academic environment, our approach was also considered by some as "too applied" to fit in such a setting. These arguments, however, are based on a number of false assumptions and preconceived ideas. One of them is that fundamental and applied research must be different entities. This account, written in an attempt to give a historical perspective on the discovery process that has led to a new family of anti-HIV agents and new insights into one of the important replicative enzymes, should prove the opposite. In our view, only good research needs to be aimed at, carried out with an open mind and with a constant awareness of fundamental and applied aspects. In our research philosophy this has also meant trying to create, based on a number of defined choices, the optimal conditions where so-called chance discoveries can take place.

References

Baba M, Pauwels R, Herdewijn P, De Clercq E, Desmyter J, Vandeputte M (1987): Both 2′,3′-dideoxythymidine and its 2′,3′-unsaturated derivative (2′,3′-dideoxythymidinene) are potent and selective inhibitors of human immunodeficiency virus replication *in vitro. Biochem Biophys Res Commun* 142:128–134

Baba M, Tanaka H, De Clercq E, Pauwels R, Balzarini J, Schols D, Nakashima H, Perno C, Walker B, Miyasaki T (1989): Highly specific inhibition of human immunodeficiency virus type 1 by a novel 6-substituted acyclouridine derivative. *Biochem Biophys Res Commun* 165:1375–1381

Baba M, De Clercq, E, Tanaka, H, Ubasawa, M, Takashima, H, Sekiya, K, Nitta, I, Umezu, K, Nakashima, H, More, S, Shigeta, S, Walker, RT, Miyasaka, T (1991): Potent and selective inhibition of human immunodeficiency virus type 1 (HIV-1 by 5-ethyl-6-phenylthiouracil derivatives through interaction with the HIV-1 reverse transcriptase. *Proc Natl Acad Sci USA* 88:2356–2360

Broder S, Yarchoan R, Collins JM, Malen HC, Markham PD, Klecker RW, Redfield RR, Mitsuya H, Hoth DF, Gelmann E, Groopman JE, Resnick L, Gallo RC, Myers CE, Fauci AS (1985): Effects of suramin on HTLV-III/LAV infection presenting as Kaposi's sarcoma or AIDS-related complex: Clinical pharmacology and suppression of virus replication in vivo. *Lancet* 2:627–630

Clavel F, Guayader M, Guétard D, Brun-Vézinet F, Chamaret S, Rey MA, Santos-Ferreira O, Laurent AG, Dauguet C, Katlama C, Rouzioux C, Klatzmann D, Champalimaud JL, Montagnier L (1986): Isolation of a new human retrovirus from West African patients with AIDS. *Science* 233:343–346

Debyser Z, Pauwels R, Andries K, Desmyter J, Kukla M, Janssen PAJ, De Clercq E (1991): An antiviral target on reverse transcriptase of human immunodeficiency virus type 1 revealed by tetrahydroimidazo[4,5,1-*jk*][1,4]-benzodiazepin-2(1*H*)-one and -thione (TIBO) derivatives. *Proc Natl Acad Sci USA* 88:1541–1455

Debyser Z, Pauwels R, Andries K, Desmyter J, Engelborghs Y, Janssen PAJ, and De Clercq E (1992a): Allosteric inhibition of HIV-1 reverse transcriptase by TIBO compounds. *Mol Pharmacol* 41:203–208

Debyser Z, De Vreese K, Pauwels R, Yamamoto N, Anné J, De Clercq E, Desmyter J. (1992b): Differential inhibitory effects of TIBO derivatives on different strains of simian immunodeficiency virus. *J Gen Virol* 73:1799–1804

De Clercq E (1979): Suramin a potent inhibitor of reverse transcriptases. *Cancer Lett* 8:9–22

De Vreese K, Debyser Z, Pauwels R, Desmyter J, De Clercq E, Anné J, (1992): Resistance of human immunodeficiency virus type 1 reverse transcriptase to TIBO derivatives induced by site-directed mutagenesis. *Virology* 188:900–904

Fischl MA, Richman DD, Grieco MH, Gottlieb MS, Volberding PA, Laskin OL, Leedom JM, Groopman JE, Mildvan D, Schooley RT, Jackson GG, Durack DT, King D, AZT Collaborative Working Group (1987): The efficacy of azidothymidine (AZT) in the treatment of patients with AIDS and AIDS-related complex. *N Engl J Med* 317:185–191

Goldman ME, Nunberg JH, O'Brien JA, Quintero JC, Schleif WA, Freund KF, Gaul SL, Saari WS, Wai JS, Hoffman JM, Anderson PS, Hupe DJ, Emini EA, Stern AM (1991): Pyridinone derivatives: Specific human immunodeficiency virus type 1 reverse transcriptase inhibitors with antiviral activity. *Proc Natl Acad Sci USA* 88:6863–6867

Koyanagi Y, Hanada S, Takahashi M, Uchino F, Yamamoto N (1985): Selective cytotoxicity of AIDS virus towards HTLV-I transformed cell lines. *Int J Cancer* 36:445–451

Kukla MJ, Breslin HJ, Pauwels R, Fedde CL, Miranda M, Scott MK, Sherril RG, Raeymaeckers A, Van Gelder J, Andries K, Janssen MA, De Clercq E, Janssen PAJ (1991a): Synthesis and anti-HIV-1 activity of 4,5,6,7-tetrahydro-5-methylimidazo[4,5,1-*jk*][1,4]benzodiazepin-2(1*H*)-one (TIBO) derivatives. *J Med Chem* 34:746–751

Kukla MJ, Breslin HJ, Diamond CJ, Grous PP, Ho CY, Miranda M, Rodgers JD, Sherrill RG, De Clercq E, Pauwels R, Andries K, Moens LJ, Janssen MAC, Janssen PAJ (1991b): Synthesis and anti-HIV-1 activity of 4,5,6,7-tetrahydro-5-methylimidazo[4,5,1-*jk*][1,4]benzodiazepin-2(1*H*)-one (TIBO) derivatives. Part 2. *J Med Chem* 34:3187–3197

Merluzzi VJ, Hargrave KD, Labadia M, Grozinger K, Skoog M, Wu JC, Shih C-K, Eckner K, Hattox S, Adams J, Rosenthal AS, Faanes R, Eckner RJ, Koup RA, Sullivan JL (1990): Inhibition of HIV-1 replication by a nonnucleoside reverse transcriptase inhibitor. *Science* 250:1411–1413

Mitsuya H, Popovic R, Yarchoan S, Matsushita RC, Gallo RC, Broder S (1984): Suramin protection of T-cells in vitro against infectivity and cytopathic effect of HTLV-III. *Science* 226:172–174

Nunberg, JH, Schleif WA, Boots EJ, O'Brien JA, Quintero JC, Hoffman JM, Emini EA, Goldman M (1991): Viral resistance to human immunodeficiency virus type 1-specific pyridinone reverse transcriptase inhibitors. *J Virol* 65:4887–4892

Pauwels R, De Clercq E, Desmyter J, Balzarini J, Goubau P, Herdewijn P, Vanderhaeghe H, Vandeputte M (1987): Sensitive and rapid assay in MT-4 cells for the detection of antiviral compounds against the AIDS virus. *J Virol Methods* 16:171–185

Pauwels R, Balzarini J, Baba M, Snoeck R, Schols D, Herdewijn P, Desmyter J, De Clercq E (1988a): Rapid and automated tetrazolium-based colorimetric assay for the detection of anti-HIV compound. *J Virol Methods* 20:309–321

Pauwels R, Balzarini J, Schols D, Baba M, Desmyter J, Rosenberg I, Holy A, De Clercq E (1988b): Phosphonylmethoxyethyl purine derivatives: A new class of anti-HIV agents. *Antimicrob Agents Chemother* 32:1025–1030

Pauwels R, Andries K, Desmyter J, Schols D, Kukla JM, Breslin HJ, Raeymaeckers A, Van Gelder J, Woestenborghs R, Heykants J, Schellekens K, Janssen MAC, De Clercq E, Janssen PAJ (1990a): Potent and selective inhibition of HIV-1 replication *in vitro* by a novel series of TIBO derivatives. *Nature (London)* 343:470–474

Pauwels R, Andries K, Desmyter J, Kukla JM, Heykants J, De Clercq E, Janssen PAJ (1990b): Potent and selective inhibition of HIV-1 replication *in vitro* by a novel series of tetrahydroimidazo[4,5,1-*jk*][1,4]-benzodiazepin-2(1*H*)-one and -thione (TIBO) derivatives. In: *Design of Anti-AIDS Drugs*, De Clercq E., ed. Amsterdam: Elsevier, pp 103–122

Pauwels R, Andries K, Debyser Z, Kukla MJ, Schols D, Breslin HJ, Woestenborghs R, Desmyter J, Janssen MAC, De Clercq E, Janssen PAJ (1992): New TIBO derivatives are potent inhibitors of HIV-1 replication, and are synergistic with 2′,3′-dideoxynucleoside analogues. Submitted for publication

Pauwels R, Andries K, Debyser Z, Van Daele P, Schols D, Stoffels P, De Vreese K, Woestenborghs R, Vandamme A-M, Janssen CGM, Anné J, Cauwenbergh G, Desmyter J, Heykants J, Janssen MAC, De Clercq E, Janssen PAJ (1993): Potent and highly selective HIV-1 inhibition by a new series of α-anilino phenylacetamide (α-APA) derivatives targeted at HIV-1 reverse transcriptase. *Proc Natl Acad Sci USA* 90:1711–1715

Pialoux G, Youle M, Dupont B, Gazzard B, Cauwenbergh GFMJ, Stoffels PAM, Davies S, De Saint Martin J, Janssen PAJ (1991): Pharmacokinetics of R82913 in patients with AIDS or AIDS-related complex. *Lancet* 338:140–143

Romero DL, Busso M, Tan C-K, Reusser F, Palmer JR, Pope SM, Aristoff PA, Downey KM, So AG, Resnick L, Tarpley WG (1991): Nonnucleoside reverse transcriptase inhibitors that potently and specifically block human immuno-deficiency virus type 1 replication. *Proc Natl Acad Sci USA* 88:8806–8810

Shih C-K, Rose JM, Hansen GL, Wu JC, Bacolla A, Griffin JA (1991): Chimeric human immunodeficiency virus type 1/type 2 reverse transcriptases display reversed sensitivity to nonnucleoside analog inhibitors. *Proc Natl Acad Sci USA* 88:9878–9882

Wu JC, Warren TC, Adams J, Proudfoot J, Skiles J, Raghavan P, Perry C, Potocki I, Farina PR, Grob PM (1991): A novel dipyridodiazepinone inhibitor of HIV-1 reverse transcriptase acts through a nonsubstrate binding site. *Biochemistry* 30:2022–2026

Yarchoan R, Klecker RW, Weinhold KJ, Markham PD, Lyerly HK, Durack DT, Gelmann E, Nusinoff Lehrman S, Blum RM, Barry DW, Shearer GM, Fischl MA, Mitsuya H, Gallo RC, Collins JM, Bolognesi DP, Myers CE, Broder S (1986): Administration of 3′-azido-3′-deoxythymidine, an inhibitor of HTLV-III replication, to patients with AIDS or AIDS-related complex. *Lancet* 1:575–580

5

Discovery and Development of 2-Pyridinone HIV-1 Reverse Transcriptase Inhibitors

MARK E. GOLDMAN

Introduction

Merck's commitment to the development of antivirals and vaccines for the treatment/prevention of human immunodeficiency virus (HIV) infection began in 1986 with the formation of an AIDS Task Force. This committee examined the HIV replication cycle to determine which targets had the greatest potential to be exploited for therapeutic intervention. Teams were then formed around each target to develop a better understanding of their biology as well as to devise screening assays for identification of leads. This mechanism-based approach toward drug development was chosen over a cell culture screening approach for many reasons. First, by understanding the molecular aspects of a target, novel sites for inhibition may become evident. Second, excellent inhibitors may be missed in a cell culture assay owing to irrelevant chemical properties (i.e., chemical instability of a compound during incubation, or inability of a compound to cross the cell membrane). Finally, the cell culture screening approach was not used in an effort to reduce the chance exposure of lab personnel to HIV infection.

Reverse transcriptase (RT) was one of the targets chosen based upon its unique polymerase activity (RNA-directed DNA synthesis) as well as the demonstration that the antiviral activity of 3'-azidothymidine (AZT) was mediated by inhibition of HIV-1 RT activity (Mitsuya et al., 1985). While there were many opportunities to license newer nucleoside analogues as development candidates, the toxicities/side effects of these drugs (Goldman, 1991a) outweighed the fact that they were proven to delay HIV disease progression. Instead, efforts were directed mainly toward screening natural product extracts for the presence of selective

The Search for Antiviral Drugs
Julian Adams and Vincent J. Merluzzi, Editors
© 1993 Birkhäuser Boston

HIV-1 RT inhibitors, based upon the precedence established with other natural product-derived substances that are selective inhibitors of polymerases. For example, aphidicolin (a tetracyclic diterpenoid) and α-amanitin (a bicyclic octapeptide), both identified from fungi, were found to be selective inhibitors of cellular DNA polymerase α and RNA polymerase II, respectively (Wieland et al., 1981; Spadari et al., 1982). In addition to the natural product screening effort, compounds chosen at random from the Merck sample collection were tested routinely for inhibitory activity in the RT assay.

Many of the antiviral drug targets required large manpower efforts for chemistry, molecular biology, and enzymology. In the case of the HIV-1 protease, for example, Merck scientists proved that the enzyme was required for biological activity by mutating the active site and demonstrating that the resulting virus was not infectious (Kohl et al., 1988). Since the HIV-1 protease was relatively small, the three-dimensional structure was determined (Navia et al., 1989) in the hope that this knowledge would aid in the drug development effort. In contrast, HIV-1 RT had already been shown to be involved in viral infectivity from mutagenesis studies (Folks et al., 1986) and pharmacological studies with nucleoside analogues (Mitsuya et al., 1985). As a result, the decision was made to initially set up only a screening program for novel HIV-1 RT inhibitors, with the anticipation of additional support once novel leads were identified. To this end, the RT group was formed in Merck's Department of New Lead Pharmacology and was initially composed of two individuals. Although members of the Department of New Lead Pharmacology did not have experience in the enzyme inhibitor discovery field, several key drug leads had been identified within this department in other therapeutic areas, including cholecystokinin antagonists and oxytocin antagonists (Chang et al., 1985; Pettibone et al., 1989), making it a logical setting for drug discovery.

Design of the Screening Program

To attain the goal of identifying selective inhibitors of HIV-1 RT that, unlike the nucleoside analogues, do not require metabolic activation, a "rational" screening plan was developed. This program involved using a battery of primary and secondary assays that could rapidly exclude nonspecific inhibitors.

Based upon the early expectation that inhibitors of avian myeloblastosis virus (AMV) RT would also inhibit HIV-1 RT, initial screening assays were set up with AMV RT. While this assumption turned out to be rather naive, the assay using AMV RT (which was in plentiful supply from commercial sources) allowed us to learn the technical concepts very rapidly. Shortly after initiation of AMV RT screening, Irving Sigal and colleagues (in Merck's Department of Molecular Biology) identified a molecular clone from the HIV-1 protease program that expressed a protein with the chromatographic and immunological properties of HIV-1 RT. This clone was turned over to the RT group. Fortunately, RT was relatively easy to express and purify from this clone. The properties of the recombinant enzyme were equivalent to virus-derived, immuno-purified RT. Subsequently, expression and purification of the HIV-1 RT was optimized (see Goldman et al., 1990), so that greater yields of active enzyme could be obtained from each preparation.

The HIV-1 RT assay is based upon the ability of this polymerase to form radioactive complementary DNA from an RNA substrate annealed to a DNA primer. For screening purposes, a synthetic homopolymer composed of poly rA annealed to oligo dT_{12-18} (rA•dT) was chosen as template•primer owing to the excellent enzyme activity reported with HIV-1 RT. Using this enzyme/template•primer system, all assay conditions were optimized (Goldman et al., 1990). In the case of the divalent cation (magnesium), a fourfold greater concentration above the initial plateau concentration (3 mM) was used to reduce indirect inhibition of RT activity by chelators that may be present in some screening samples.

The secondary assays for specificity determination included calf thymus DNA polymerase α, human (HeLa cell) DNA polymerases α, β, γ, and δ, bacterial polymerases, and other retroviral polymerases. The initial goal was to identify compounds that inhibited HIV-1 RT with at least a 10-fold selectivity over DNA polymerase α. Based upon the demonstration that some toxicities of nucleoside analogues were due to inhibition of cellular DNA polymerase γ by the nucleoside analogue triphosphates (reviewed by Goldman, 1991a), the other cellular DNA polymerase assays were also assigned a high level of importance. In order to develop a better understanding of the degree of specificity of a given compound, but not necessarily to determine if a compound was of interest as a new lead, the effects on many retroviral and bacterial polymerases were determined.

Compounds that were selective, low micromolar (or better) inhibitors of HIV-1 RT were then tested for antiviral activity (efficacy) in a cell culture–based "spread" assay developed by Emilio Emini, Jack Nunberg, and William Schleif in Merck's Department of HIV Biology (Goldman et al., 1991). Typically, human MT4 T-lymphoid cells were infected with $HIV-1_{III_B}$ (multiplicity of infection = 0.01); then, 1 day later, the compound was added, and the spread of infection through the cell culture was assessed by p24 core antigen production on day 4. The end point for this assay was the concentration of inhibitor that pre-vented the spread of infection by at least 95% (CIC_{95}). This end point was chosen to reflect our goal of trying to identify antiviral agents that completely prevented virus replication. For technical reasons, the CIC_{95} was used rather than a CIC_{100}.

Early Actives

During the first 2 years of the screening program, approximately 20,000 natural product extracts and synthetic chemicals were tested for inhibitory activity in the HIV-1 RT assay. Of these, about 1% were active and required specificity determination. After this point, only a few samples were characterized in greater detail. The first selective actives from natural product sources were identified by Gino Salituro (of Merck's Department of Natural Product Chemistry) as calcium and oxalate. Once calcium was verified experimentally to be a selective inhibitor of HIV-1 RT activity compared to cellular DNA polymerase α, 500-μM EGTA was included in the standard reaction buffer of the HIV-1 RT screening assay. In addition, all selective natural product actives were tested for the presence of oxalate by means of a simple colorimetric assay.

The first novel HIV-1 RT inhibitors identified from screening were the rubromycins (Figure 1), a class of known quinone antibacterials (Goldman et al., 1990). The HIV-1 RT-inhibitory potencies (K_i values) of β- and γ-rubromycin were 0.27 and 0.13 μM, respectively. They were 9- to 19-fold weaker as inhibitors of calf thymus DNA polymerase α. Kinetic analysis demonstrated that the inhibition was reversible, competitive with respect to template•primer, and noncompetitive with respect to deoxynucleoside triphosphate. Owing to cytotoxicity at concentrations above 6 μM, these compounds could not be evaluated for antiviral activity in the range where they should have been active. Since chemical searches of the Merck sample collection for compounds related to the rubromycins did not identify agents with better properties and since chemical modifications of the rubromycins yielded compounds that were essentially inactive, this lead was dropped.

Identification of L-345,516

In the spring of 1989, L-345,516 (Figure 2) was identified as an inhibitor of HIV-1 RT ($IC_{50} = 1$ μM) by use of the rA•dT/TTP substrates. It did not significantly inhibit calf thymus DNA polymerase α at 300 μM and, therefore, met our selectivity criteria for consideration as a lead. This indomethacin analogue was first synthesized in 1972 by Bruce Witzel (Department of Synthetic Chemistry Research) for evaluation in an inflammation assay and was subsequently chosen for random screening in Merck assays by George Hartman (Department of Medicinal Chemistry) based upon his perception that it was structurally similar to certain

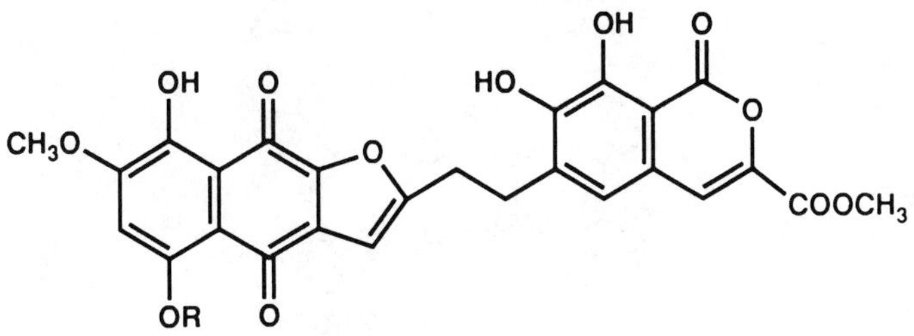

$R = CH_3$ α- Rubromycin
R = H γ-*iso*- Rubromycin

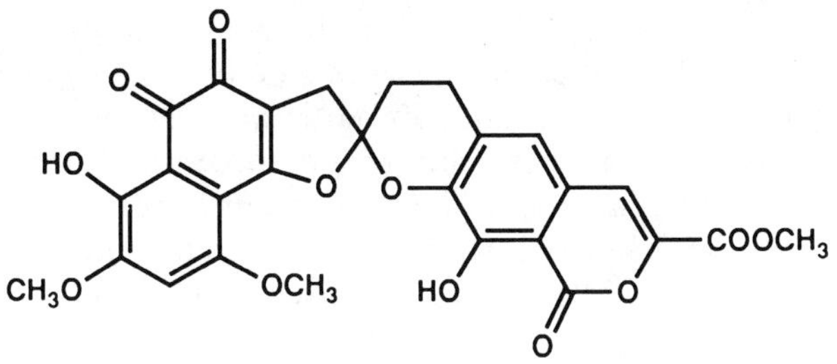

β- Rubromycin

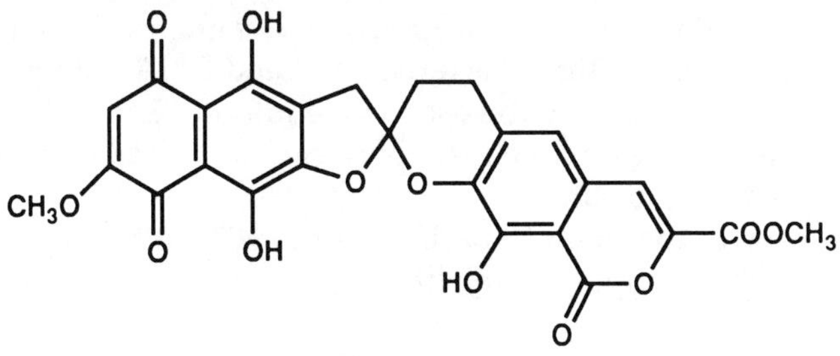

γ- Rubromycin

FIGURE 1. Structures of rubromycins.

non-benzodiazepine ligands of the anxiolytic benzodiazepine receptor. Coincidentally, two other chemical classes of selective HIV-1 RT inhibitors subsequently reported are benzodiazepine analogues (Pauwels et al., 1990; Merluzzi et al., 1990; see also Chapters 3 and 4 of this volume).

Once L-345,516 was identified as a selective lead, its HIV-1 RT-

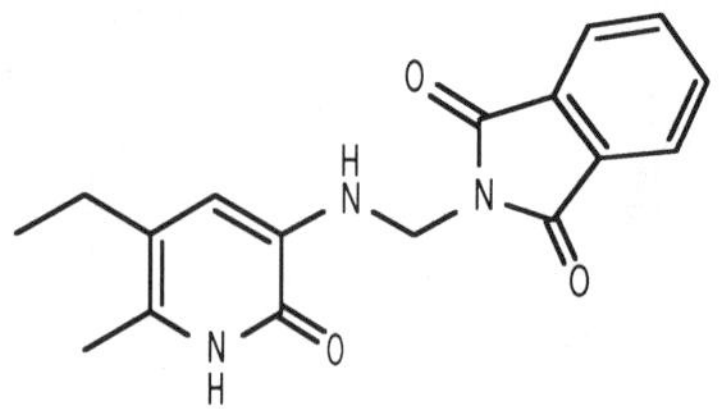

L-345,516

FIGURE 2. HIV-1 RT inhibitor screening lead.

inhibitory properties were evaluated in greater detail. Initial characterization of L-345,516 demonstrated that there was no difference in activity with Tris vs. Hepes buffers and the inhibition was independent of dithiothreitol (DTT), bovine serum albumin (BSA), or dimethyl sulfoxide (DMSO) concentration. In addition, there was a significant template•primer dependence with greater potencies using rC•dG or dA•dT than rA•dT (Table 1), a divalent cation dependence (Figure 3), and noncompetitive kinetics with respect to substrates. Based upon these findings, the hypothesis was proposed that L-345,516 inhibited in the product mode much like phosphonoformate, which also possesses a template•primer dependence, cation dependence, and noncompetitive kinetics with respect to substrates (Oberg, 1989).

We were encouraged with the greater potency of L-345,516 for HIV-1 RT over calf thymus DNA polymerase α and characterized the selectivity in greater depth. Although L-345,516 had an IC_{50} of 0.03 μM on HIV-1 RT, at 300 μM (10,000-fold higher) it did not significantly inhibit five other viral polymerases, three bacterial RNA and DNA polymerases, or four mammalian DNA polymerases. In addition, L-345,516 did not significantly inhibit the ribonuclease H activity associated with HIV-1 RT using [^{3}H]rG•dC as substrate.

Excitement over the unique specificity of L-345,516 rapidly turned to concern that the compound might only be a selective inhibitor of

TABLE 1. Potencies of specific HIV-1 RT inhibitors: template•primer dependence

	HIV-1 RT IC_{50} (nM)				
Compound#	rC•dG	dA•dT	rA•dT	rRNA•15 mer	Globin mRNA•dT
L-345,516	30	32	1000	ND*	ND*
L-696,229	18	24	501	21	550
L-697,639	20	20	600	28	240
L-697,661	19	30	830	38	510

*Not determined.

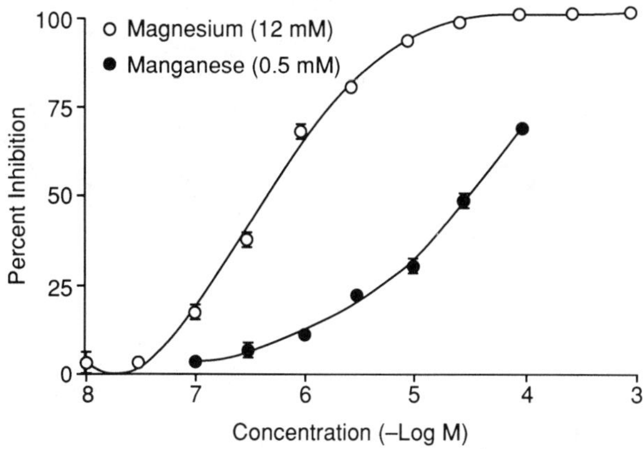

FIGURE 3. Inhibition of HIV-1 RT by L-345,516: comparison of divalent cations.

recombinant HIV-1 RT. This fear was allayed by the demonstration of equipotent inhibitory activity on virus-derived, immunopurified HIV-1 RT as well as RT-inhibitory activity on Triton X-100 lysates of four HIV-1 laboratory isolates (RUTZ, WMJ, III$_B$, and MN). These data suggested that L-345,516 interacted with a site present solely on HIV-1 RT. The concern remained, however, that the virus would be able to mutate around compounds in this pharmacological class. Although screening often ends when leads are identified, we have expanded screening efforts and laboratory personnel during the subsequent 2.5 years since discovery of L-345,516 in the hope of finding other pharmacological classes of selective HIV-1 RT inhibitors.

Soon after its discovery as an RT inhibitor, Joseph Vacca (of Merck's Department of Medicinal Chemistry) suggested that L-345,516 might be unstable in aqueous solutions owing to hydrolysis (Figure 4) of the aminal functionality. This concern was confirmed experimentally with the demonstration that the potency of L-345,516 in the RT assay decreased in a time-dependent manner when it was first diluted in aqueous solutions prior to being added to the RT assay (Figure 5). By means of HPLC (high-performance liquid chromatography), the half-life in water was found to be approximately 120 min.

Prior to the initiation of medicinal chemistry efforts on this project, it was important to demonstrate that the inhibitory activity was not a function of its decomposition. While the breakdown products (aminopyridinone, formaldehyde, phthalimide) were inactive as RT inhibitors, it was possible that nonspecific modifications to either the substrates or enzyme could have been occurring *during* the decomposition process. An experiment proposed by Andrew Stern (Department of Biological Chemistry) was carried out to address this issue. The ability of L-345,516

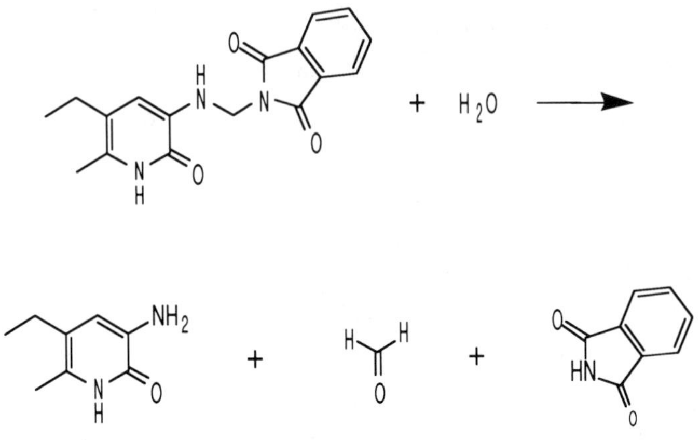

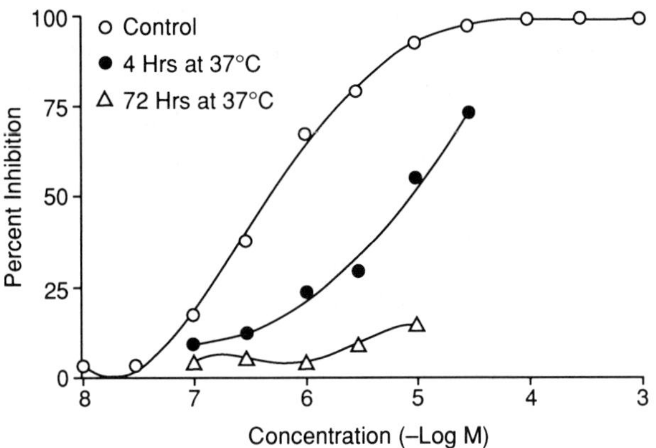

FIGURE 4. L-345,516 is unstable in aqueous solutions.

FIGURE 5. Aqueous stability of L-345,516: time dependence.

(100 μM) to inhibit reactions containing AMV RT (which is not directly inhibited by L-345,516), HIV-1 RT, or several combinations of AMV RT plus HIV-1 RT in different ratios were determined. If HIV-1 RT–mediated hydrolysis of L-345,516 was occurring, it was possible that the reactive species would interact with either template•primer or AMV RT as well as HIV-1 RT. The results of this experiment (Figure 6) demonstrated that the inhibitory activity of L-345,516 was a function of the HIV-1 RT concentration and independent of the presence of AMV RT. These results suggested that the breakdown of L-345,516 did not generally inhibit polymerase activity or destroy the substrates.

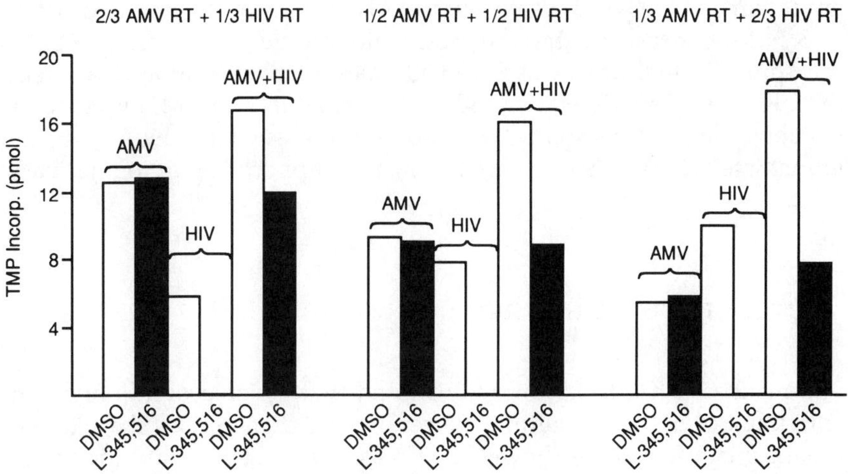

FIGURE 6. L-345,516 selectively inhibits HIV-1 RT activity.

Initiation of Synthetic Chemistry

Based upon the conclusion that L-345,516 was interacting specifically and reversibly, a small synthetic program was initiated. As first proposed by Joe Vacca, replacement of the NH in the linker with methylene was a means of eliminating the hydrolytic instability of L-345,516. This compound, L-693,593 (Figure 7), was synthesized by Jacob Hoffman (Department of Medicinal Chemistry) and was

	X=	Potency (nM)	
		IC$_{50}$	CIC$_{95}$
L-345,516	NH	30	>>50,000
L-693,593	CH$_2$	3,700	40,000

FIGURE 7. Effect of linker substitution on activity.

approximately 100-fold weaker than L-345,516 as an RT inhibitor (IC_{50} = 3.7 μM) but showed the same selectivity profile (Saari et al., 1991). In the spread assay, however, L-693,593 had a CIC_{95} of 40 μM, whereas the CIC_{95} of L-345,516 was significantly greater than 50 μM, owing to its instability in aqueous solutions. Based upon these data, there was a an immediate 10-fold increase in chemistry support beginning January 1990.

The Chemistry Program

The results with L-693,593 demonstrated that a stable pyridinone had antiviral activity in cell culture. Since this one change caused a 100-fold loss of intrinsic activity, however, it was clear that a significant chemistry effort would be necessary to turn these inhibitors into drugs. Under the direction of Paul Anderson and Walfred Saari (Department of Medicinal Chemistry), the initial goal of the chemistry program was to increase potency by first determining the optimal substitutions on the pyridinone and the appropriate linkers between the two heterocycles then finding a replacement for the phthalimide.

In the aminomethyl series, a number of substitutions were made on the pyridinone ring (Table 2; see Saari et al., 1991). Removal of the alkyl groups in the 5 and 6 positions (compound 1) or just the 5 position (compound 2) caused a significant loss of RT-inhibitory activity. The 5,6-dimethyl analogue (compound 3) was also weaker. Switching the

TABLE 2. Pyridinone substitutions in the phthalimide series

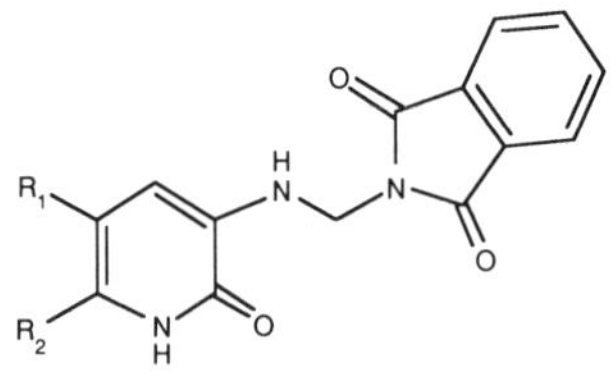

Compound#	R_1	R_2	IC_{50}(nM)
L-345,516	$-CH_2CH_3$	$-CH_3$	30
1	H	H	800
2	H	$-CH_3$	9,000
3	$-CH_3$	$-CH_3$	250
4	$-CH_3$	$-CH_2CH_3$	1,000
5	$-CH_2CH_2CH_3$	$-CH_3$	31
6	$-CH_2CH_2CH_2CH_3$	$-CH_3$	61

methyl and ethyl groups of L-345,516 (compound 4) resulted in a loss of activity and further highlighted the lack of tolerance in this region. In the presence of the 6-methyl group, increasing the length of the alkyl chain did not significantly change activity (compounds 5 and 6). Many other substitutions in the 4, 5, and 6 positions were examined but did not result in better compounds (Hoffman et al., 1993; Saari et al., 1993).

Using the 5-ethyl-6-methyl-2-pyridinone in the phthalimide series, the Merck scientists synthesized and evaluated many linker substitutions (Table 3). In the aminoalkyl class, aminomethyl (L-345,516) was significantly better than aminoethyl (compound 7) or amino-*n*-propyl (compound 8). Similarly, in the carba class, the ethylene linker (L-693,593) was significantly better than shorter (compound 9) or longer (compound 10) linkers or the ethynyl linker (trans configuration, compound 11).

Based upon the conclusion that the 5-ethyl-6-methyl-2-pyridinone was optimal, replacements for the phthalimide were made in both the aminomethyl and ethylene series (Table 4). The aminomethyl series was pursued in Walfred Saari's lab, where all phthalimide substitutions were less potent than L-345,516. Of these, however, benzoxazole (L-696,040) was better than all others, including benzimidazole (compound 12), benzyl (compound 13), and naphthyl (compound 14).

Since the ethylene linker analogue in the phthalimide series (L-693,593) was 100-fold weaker than the equivalent compound in the aminomethyl series (L-345,516), the ethylene linker analogue in the benzoxazole series was not expected to be very active. In spite of this

TABLE 3. Linker substitutions in the phthalimide series

Compound#	Y	IC_{50} (nM)
L-345, 516	$-NHCH_2-$	30
7	$-NHCH_2CH_2-$	45,000
8	$-NHCH_2CH_2CH_2-$	190,000
9	$-CH_2-$	26,500
L-693,593	$-CH_2CH_2-$	3,700
10	$-CH_2CH_2CH_2-$	210,000
11	$-CHCH-$ (trans)	>300,000

logic (from a biologist's view), Jacob Hoffman synthesized the compound (L-696,229) anyway. L-696,229 turned out to be 10-fold *more* potent than L-696,040 and even more potent than the original lead, L-345,516 (Table 4).

In the hope of improving the potency of the L-696,040 series, nuclear substitutions on the benzoxazole were explored. The results demonstrated that methyl or chlorine substitutions in the 4 or 7 positions (compounds 15, 18, 19, and 20; see Table 5) improved potency whereas substitutions in the 5 or 6 positions (compounds 16 and 17) reduced potency. Simultaneous substitutions of methyl groups or chlorine atoms in the 4 and 7 positions (L-697,639 or L-697,661, respectively) significantly improved RT-inhibitory activity compared to L-696,040. In contrast to the improvements with nuclear substitutions in the aminomethyl series, there was not a great improvement in the ethylene linker series (Saari et al., 1991; Hoffman et al., 1993), giving further evidence for different structure–activity profiles between the two classes.

From this 4-month chemistry effort, three highly potent compounds (L-696,229, L-697,639, and L-697,661) were identified. These compounds

TABLE 4. Aromatic/heterocyclic group replacements

Compound#	X=	R=	IC$_{50}$ (nM)
L-345,516	NH		30
12	NH		6,400
13	NH		5,300
14	NH		440
L-696,040	NH		210
L-696,229	CH$_2$		18

TABLE 5. Benzoxazole ring substitutions in the aminomethyl linker series

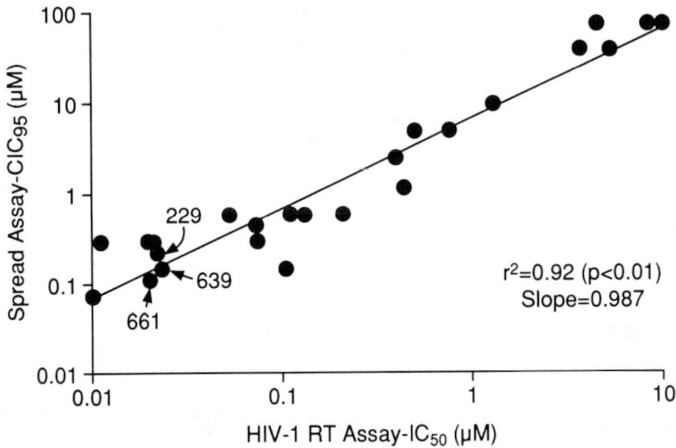

Compound#	Substituent(s)	IC_{50} (nM)
L-696,040	–	210
15	4-Me	125
16	5-Me	1,370
17	6-Me	1,700
18	7-Me	55
L-697,639	4,7-Me$_2$	19
19	4-Cl	53
20	7-Cl	105
L-697,661	4,7-Cl$_2$	19

were stable in physiological solutions and demonstrated specificity for HIV-1 RT over all other enzymes examined. There was a close correlation between the abilities of representative compounds in the pyridinone class to inhibit RT activity and spread of HIV-1 infection in cell culture (Figure 8), suggesting that RT was their target for antiviral activity and that metabolic activation of the compounds by target cells was not required for biological activity (Goldman et al., 1991).

FIGURE 8. Correlation of enzyme and spread potencies.

Biological Properties of Lead Compounds

L-696,229, L-697,639, and L-697,661 each inhibited HIV-1 RT activity with similar potencies (Table 1). As found with the original lead, L-345,516, their potencies were all better with rC•dG or dA•dT as template•primer than with rA•dT (Table 1; Goldman et al., 1991, 1992). In addition, all three compounds were noncompetitive inhibitors with respect to deoxynucleoside triphosphates and were more potent in the presence of magnesium than manganese.

As illustrated with homopolymer substrates, we also found a divergence of potencies with various heteropolymer substrates (Table 1). The globin mRNA substrate, originally developed for evaluation of antisense compounds (Byrn et al., 1991), did not yield high potency with the specific HIV-1 RT inhibitors. When a ribosomal RNA assay was used (White et al., 1991), however, the compounds had potencies close to the rC•dG and dA•dT potencies. These results suggest that the range of potencies for this class of compounds are not a function of the composition of the *synthetic* substrates.

Comprehensive studies in the laboratory of Andrew Stern yielded further insight into the mechanism of RT inhibition by these compounds (Goldman et al., 1991). The inhibition was reversible; after incubation of compound with HIV-1 RT and substrates, removal of the compound (G-25 chromatography or dilution) resulted in a restoration of enzyme activity. By HPLC, unchanged compound was also recovered quantitatively. Binding studies with [^{3}H]L-697,639 and the rC•dG/dGTP system demonstrated that high-affinity binding, under the conditions of this assay, was dependent upon the presence of both substrates, magnesium, and enzyme, with approximately 1 mol of inhibitor bound per mol of heterodimeric RT. These results gave further support to the hypothesis that these compounds interact with enzyme in a noncompetitive, magnesium-dependent manner. Based upon the indirect evidence presented earlier that these compounds act in the product mode much like phosphonoformate (PFA), the ability of PFA to interact with enzyme•substrate•inhibitor complexes was examined. The results demonstrated that PFA could both prevent association of enzyme•substrate•inhibitor complexes as well as cause dissociation of preformed complexes.

Antiviral Activity of Lead Compounds

Evaluation of the antiviral activity in the cell culture spread assay required the dedicated efforts of William Schleif, Julio Quintero, and (more recently) Audrey Rhodes and Donna Titus (of Merck's Depart-

TABLE 6. Antiviral properties of L-696,229, L-697,639, and L-697,661 in cell culture: comparison with AZT

HIV-1	Cell Type	$CIC_{95}{}^+$ (nM)			
		L-696,229	L-697,639	L-697,661	AZT
III_B	H9	200	100	200	20,000
III_B	MT4	100	25-50	25-50	12-25
III_B	PBMC*	100	50	100	200
MN	MT4	100	100	50	100
WMJ-2	MT4	100	200	100	25
RF	MT4	50	50	50	ND**
RUTZ	MT4	50	100	50	ND**
A018A/H112-2	CEM	100	25	25	12
A018c/G910-6	CEM	50	12	12	1600
rA17	MT4	100,000	50,000	50,000	25

$^+CIC_{95}$ = cell culture inhibitor concentration that prevented, by at least 95%, the spread of HIV-1 infection.
*Human peripheral blood mononuclear cells.
**Not determined.

ment of HIV Biology). Their studies demonstrated that the three lead compounds prevented the spread of HIV-1 infection (Table 6; see Goldman et al., 1991, 1992; Nunberg et al., 1991). With the exception of the rA17 viral isolate, the three compounds had CIC_{95} potencies of 0.20 μM or less on all laboratory isolates. The rA17 isolate is a variant of HIV-1$_{III_B}$ selected for resistance to the pyridinone class of specific HIV-1 RT inhibitors (Nunberg et al., 1991; see the *Resistance* section, below). These compounds also prevented the spread of HIV-1$_{III_B}$ infection in phytohemagglutinin-stimulated peripheral blood mononuclear cells maintained in a medium containing interleukin-2.

L-696,229, L-697,639, and L-697,661 were also tested in the spread assay using the viral isolates A018A/H112–2 and A018C/G910–6, which were obtained, respectively, from an HIV-1 infected person before the initiation of AZT therapy and after 6 months of AZT therapy (Larder et al., 1989). The three compounds possessed similar antiviral activity to that of the two isolates (Table 6), suggesting no cross-resistance to this class of compounds following chronic AZT administration. In contrast, AZT was significantly weaker on the A018C/G910–6 isolate than on the A018A/H112–2 isolate (Table 6).

Synergy with Nucleoside Analogues in Cell Culture

Combination therapy is often used for the treatment of bacterial or viral infections. For both types of infections, combinations have the ability, if used properly, to increase efficacy and delay or prevent the development of drug-resistant variants. Since AZT is the standard of care for treating HIV-1 infection, demonstration that the specific HIV-1 RT inhibitors did not antagonize the effects of AZT was also important.

In the spread assay, the antiviral activity of the pyridinones was determined alone or in combination with nucleoside analogues using the fractional inhibitory concentration method (Goldman et al., 1991). The results demonstrated that combinations of L-697,639 or L-697,661 with AZT or dideoxyinosine resulted in greater activity compared to any of the agents alone (Figure 9; see Goldman et al., 1991). These results suggested that combinations of specific HIV-1 RT inhibitors with nucleoside analogues, which do not have common toxicities/side effects, may increase the effectiveness of antiviral activity and possibly delay the appearance of drug-resistant variants.

Resistance to Specific HIV-1 RT Inhibitors

The only good statement that can be made about resistance is that it proves a drug treatment *was* effective. In other words, resistance should not reproducibly occur if there is no selective pressure on the system to generate the drug-resistant variants.

Resistance was a concern within Merck from the initial demonstration that L-345,516 was highly specific for HIV-1 RT over all other enzymes examined and that it was not a competitive inhibitor with respect to substrates. Our plan was to attempt to generate drug-resistant HIV-1 variants in cell culture once hydrolytically stable analogues were synthesized. It was possible that by understanding resistance to these compounds we might be able to develop a second generation of RT

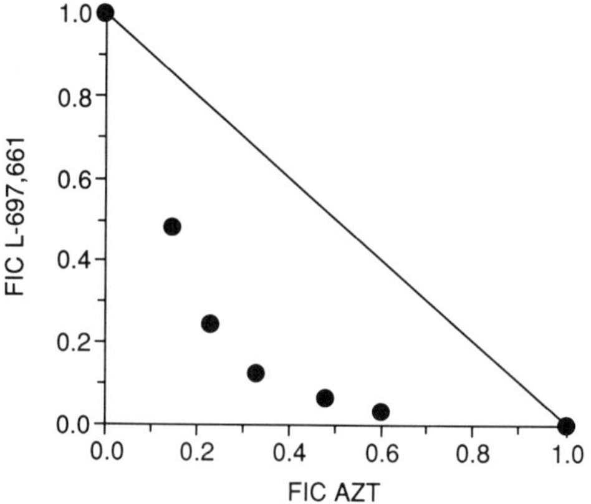

FIGURE 9. Synergy between L-697,661 and AZT.

inhibitors active on resistant virus or at least design appropriate clinical studies.

Under the direction of Jack Nunberg and Emilio Emini (Department of HIV Biology), the study was specifically designed to balance the needs for multiple rounds of viral replication and genetic diversity in each inoculum (Nunberg et al., 1991). H9 human T-lymphoid cells were incubated with HIV-1$_{III_B}$ in the presence of 40-μM L-693,593. One day later, medium and cells were separated so that free virus could be discarded. The cells were then resuspended in fresh medium containing 40-μM L-693,593. When 100% of MT4 cells from this first culture were infected, an aliquot of medium (containing virus produced by the infected cells) was added to fresh cultures in the presence of 100-μM L-693,593. After a total of four passages, cells were cloned out using FDA-H9 cells. HIV-1 isolates from the cellular clones were then characterized in standard MT4 cell spread assays using a p24 core antigen ELISA (enzyme-linked immunosorbent assay). The results demonstrated that by passage 4, there was a clear growth advantage of the selected virus propogated in the presence of L-693,593. The cellular clones A17 and B17 both showed approximately 1000-fold resistance to pyridinones including L-696,229, L-697,639, and L-697,661 but not to AZT or PFA.

Prior to determination of the RT gene DNA sequences from the resistant viruses, the RT gene from each biological clone was expressed as a fusion protein in *Escherichia coli.* In addition, specific regions of the L-693,593-resistant RT genes were subcloned into wild-type RT expression plasmids. All RTs expressed in this system possessed polymerase activity. RTs from the L-693,593-resistant virus containing either of two distinct restriction fragments were partially resistant to L-697,639 (Figure 10) but not to AZT or PFA. DNA sequencing of the resistant RT gene revealed that each of the fragments responsible for partial resistance contained a single base mutation resulting in a different amino acid (Nunberg et al., 1991). The mutation in the BamH1 to EcoRV region was a lysine to asparagine at amino acid 103 (K103N). The mutation in the EcoRV to PflM1 region was tyrosine to cysteine (Y181C). Based upon RT assays with the fragments, the K103N mutation was responsible for approximately 20-fold resistance while the Y181C mutation was responsible for approximately 800-fold resistance. Together, the two mutations were more than additive, yielding 2000-fold resistance.

Verification that these two single base changes were responsible for resistance to the pyridinone RT inhibitors was derived by constructing recombinant virus (rA17 from the cellular clone A17) that contained only these two changes. The rA17 virus showed an equivalent pharmacological profile to the A17 virus (Nunberg et al., 1991).

The mutations identified in this resistance study were found to be located in regions important for enzyme activity (Nunberg et al., 1991). Both amino acids are invariant in all sequenced HIV-1 isolates. The

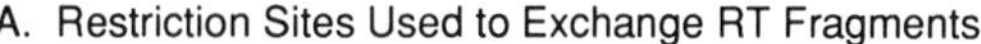

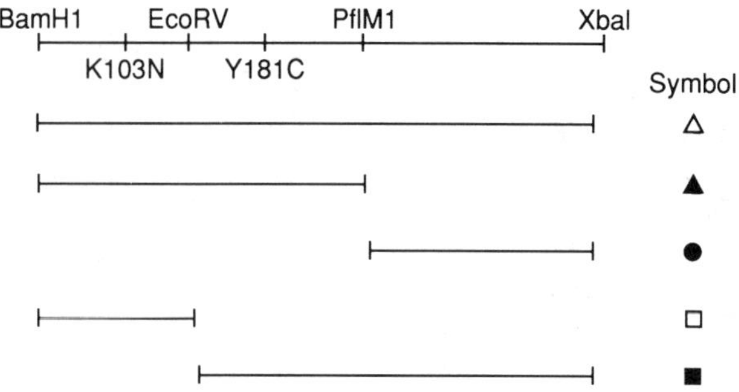

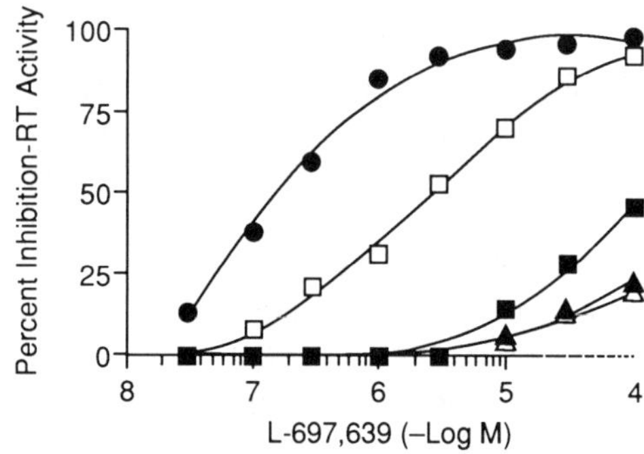

FIGURE 10. Inhibition of bacterially expressed RTs by L-697,639.

K103N mutation lies adjacent to a region that has been implicated in pyrophosphate binding (Larder et al., 1987), giving further support to our hypothesis that these compounds interact at an allosteric site related to PFA. The Y181C mutation lies near the YG/MDD motif identified in most polymerases.

Pharmacokinetics

Once potent/effective inhibitors were identified, the evaluation of their pharmacokinetic properties in vivo was necessary. This issue was especially important for the RT inhibitor project since an *acceptable* animal model for efficacy did not exist. To quickly gain insight into the absorption and metabolism of these compounds, a high-volume bioassay system for estimation of parent compound and active metabolites was set up (O'Brien et al., 1992). Plasma from animals receiving compound or control plasma spiked with known concentrations of compound were extracted with solvent. The dried extracts were dissolved in DMSO, and aliquots were then added to RT assays. The amount of inhibitory activity in experimental samples was estimated from a standard curve of spiked samples. The levels of parent compound from the same solvent extracts were subsequently determined by HPLC.

When this bioassay was used, the intravenous half-life values in rhesus monkeys were 24, 120, and 30 min for L-696,229, L-697,639, and L-697,661, respectively. Following oral administration of a clinically acceptable formulation (Drazen Ostovic, Department of Pharmaceutical Research and Development) to rhesus monkeys, peak plasma levels were greatest with L-696,229 and L-697,639. Both L-697,639 and L-697,661 had longer apparent oral half-life values than that of L-696,229.

In vitro plasma protein binding studies were carried out using normal human plasma. Of the three lead pyridinones, L-696,229 was the least protein bound (Halpin et al., 1992).

Development

Some 3–4 months after initial synthesis of the lead compounds, they were approved for development and a Reverse Transcriptase Project Team was formed. The purpose of the team was to coordinate Merck-wide developmental activities, including preclinical safety evaluation, formulations, clinical studies, human pharmacokinetics/drug metabolism, and drug supplies.

A L-697,639 Clinical Studies Application was filed in Belgium in October 1990. Under the direction of Oscar Laskin (Department of Clinical Pharmacology), Phase I studies in HIV-1 seronegative individuals began shortly thereafter. This initial clinical study started 7 months after L-697,639 was first synthesized, a Merck record. In the United States, an Investigational New Drug (IND) application was filed in November 1990 and single-dose studies in HIV-1 seropositive individuals began in December 1990 at the National Institutes of Health (NIH). These single-dose studies were double-blind and placebo-controlled. The doses for both studies were 25-, 50-, 100-, 200-, and 500-mg capsules and a 200-mg suspension. L-697, 639 was well tolerated, and there were no important changes in clinical or laboratory parameters that were clearly drug related (Davey et al., 1991; Laskin et al., 1991). The mean peak plasma levels of parent compound at the 500-mg dose was 4.2 μM (120 min).

Similar Phase I studies were conducted with L-697,661 [capsules only

(Davey et al., 1991; Laskin et al., 1991)]. As found with L-697,639, no clinically important adverse events were noted. At a dose of 500 mg to seronegative individuals, the mean peak plasma level was 1.8 μM (100 min). A 10-day multiple-dose study (100 mg q. 12 h.; 100 mg q. 8 h.; 500 mg q. 24 h.) was undertaken with similar results and no clear indication of enzyme induction or drug accumulation (Laskin et al., 1991; Goldman, 1991b).

Based upon clinical results and in vitro studies, L-697,661 was chosen as the lead compound for Phase IIa evaluation of safety/tolerability, pharmacokinetics, and pharmacodynamics in the United States and Europe (Davey et al., 1991; Goldman, 1991b). The doses of L-697,661 used were 25 mg q. 12 h., 100 mg q. 8 h., and 500 mg q. 12 h. Depending upon the investigational site, the controls were placebo and/or AZT. The study duration was 6 weeks, with a 6-week extension. The main biological activity markers examined were serum p24 core antigenemia, CD4 counts, and plasma viremia.

Although it was synthesized one month before L-697,639/L-697,661, initially we were not as eager to develop L-696,229 as the other compounds owing to its short intravenous half-life in animals. Based upon the demonstration that the protein binding with this compound was less than that of the aminopyridinones, support for development of L-696,229 grew. In May 1991 an IND application was filed for this compound. Phase I clinical studies were initiated several weeks later in HIV-1 seronegative individuals.

Perspective

Using different drug discovery approaches, at least three groups of researchers have identified chemically distinct members of the same (or a very similar) pharmacological class (Pauwels et al., 1990; Merluzzi et al., 1990; Goldman et al., 1991; see also Chapters 3 and 4 in this volume). All three groups probably hypothesized quite early that the mechanism and specificity of these drugs could potentially result in rapid induction of drug resistance in vivo. While this hypothesis may have led others investigators away from developing compounds in this pharmacological class, several were willing to take this risk in the hopes of helping those infected with this devastating virus. By understanding the nature of the interaction of these compounds with HIV-1 RT, it may be possible in the future to develop drugs with a broader spectrum of action and possibly overcome the only problem known to be associated with this first generation of specific HIV-1 RT inhibitors.

Acknowledgments

At present, more than 100 people have made important contributions to bring the project to the point that it is at today. I ask that the readers direct their attention to the original research articles cited in this review.

I would like to give special recognition to Julie O'Brien, my colleague and friend, for her great efforts and sacrifices, and to Joe Vacca, who made many excellent suggestions although he was not directly involved in the project.

References

Byrn SR, Carlson DV, Chen JK, Cushman MS, Goldman ME, Ma WP, Pidgeon CL, Ray, KA, Stowell JG, Weith HL (1991): Drug-oligonucleotide conjugates. *Adv Drug Delivery Rev* 6:287–308

Chang RSL, Lotti VJ, Monaghan, RL, Birnbaum J, Stapley, EO, Goetz MA, Albers-Schonberg G, Patchett AA, Liesch JM, Hensens, OD, Springer JP (1985): A potent nonpeptide cholecystokinin antagonist selective for peripheral tissues isolated from *Aspergillus alliadus*. *Science* 230:177–179

Davey, R Jr, Laskin O, Decker M, O'Neill D, Haneiwich S, Metcalf J, Polis M, Kovacs J, Davis S, Mauer M, Yoder C, Patterson P, Justice S, Yeh KC, Woolf E, Au T, Lane HC (1991): L-697,639 and L-697,661: Novel agents for treatment of human immunodeficiency virus type 1 infection. *31st Intersci Conf Antimicrob Agents Chemother.*, Abstr 697

Folks TM, Powell D, Lightfoote M, Koenig S, Fauci A, Benn S, Rabson A, Dagherty D, Gendelman HE, Hoggan MD, Venkatesan S, Martin MA (1986): Biological and biochemical characterization of a cloned leu-3-cell surviving infection with the AIDS virus. *J Exp Med* 164:280–290

Goldman ME (1991a): Molecular strategies for the inhibition of human immunodeficiency virus reverse transcriptase. *Am J Pharm Educ* 55:387–390

Goldman ME (1991b): L-697,661. *AIDS Clin Trial Groups Primary Infect Comm Meet, Washington DC, 1991*

Goldman ME, Salituro GS, Bowen JA, Williamson JM, Zink DL, Schleif WA, Emini EA (1990): Inhibition of human immunodeficiency virus-1 reverse transcriptase activity by rubromycins: Competitive interaction at the template•primer site. *Mol Pharmacol* 38:20–25

Goldman ME, Nunberg JH, O'Brien JA, Quintero JC, Schleif WA, Freund KF, Gaul SL, Saari WS, Wai JS, Hoffman JM, Anderson PS, Hupe DJ, Emini EA, Stern AM (1991): Pyridinone derivatives: Specific human immunodeficiency virus type 1 reverse transcriptase inhibitors with antiviral activity. *Proc Natl Acad Sci USA* 88:6863–6867

Goldman ME, O'Brien JA, Ruffing TL, Nunberg JH, Schleif WA, Quintero JC, Siegl PKS, Hoffman JM, Smith AM, Emini EA (1992): L-696,229 specifically inhibits HIV-1 reverse transcriptase and possesses antiviral activity, *in vitro*. *Antimicrob Agents Chemother* 36:1019–1023

Halpin RA, Geer LA, Wyatt DW, Vyas KP, Hichens MP (1992): The disposition of L-696,229, a potent and specific inhibitor of human immunodeficiency virus reverse transcriptase in rats and Rhesus monkeys. *FASEB J.*, in press

Hoffman JM, Wai JS, Thomas CM, Levin RB, O'Brien JA, Goldman, ME (1992): Synthesis and evaluation of 2-pyridone derivatives as HIV-1 specific reverse transcriptase inhibitors. 1. Phthalimidoalkyl and -alkylamino analogues. *J Med Chem* 35:3784–3791

Kohl NE, Emini EA, Schleif WA, Davis LJ, Heimbach JC, Dixon RAF, Scolnick EM, Sigal IS (1988): Active human immunodeficiency virus protease is required for viral infectivity. *Proc Natl Acad Sci USA* 85:4696–4690

Larder BA, Purifoy DJM, Powell KL, Darby G (1987): Site-specific mutagenesis of AIDS virus reverse transcriptase. *Nature (London)* 327:716–717

Larder BA, Darby G, Richman DD (1989): HIV with reduced sensitivity to zidovudine (AZT) isolated during prolonged therapy. *Science* 243:1731–1734

Laskin, OL, Dupont AG, Buntinx A, Schoors D, De Pré M, Van Hecken, A, Yeh KC, Woolf E, Eisenhandler R, De Smet M, Patterson P, De Lepeleire I, De Schepper PJ (1991): Human pharmacokinetics and tolerability of L-697,661 and L-697,639: Nonnucleoside human immunodeficiency virus type 1 reverse transcriptase inhibitors. *31st Intersci Conf Antimicrob Agents Chemother*, Abstr 698

Merluzzi VJ, Hargrave KD, Labadia M, Grozinger K, Skoog M, Wu JC, Shih K-C, Eckner K, Hattox S, Adams J, Rosenthal AS, Faanes R, Eckner RJ, Koup RA, Sullivan JL (1990): Inhibition of HIV-1 replication by a nonnucleoside reverse transcriptase inhibitor. *Science* 250:1411–1413

Mitsuya H, Weinhold KJ, Furman PJ, St Clair MII, Lehrman SN, Gallo RC, Bolognesi D, Barry DW, Broder S (1985): 3'-Azido-3'-deoxythymidine (BW A509U): An antiviral agent that inhibits the infectivity and cytopathic effect of human T-lymphotropic virus type III/lymphadenopathy-associated virus *in vitro. Proc Natl Acad Sci USA* 82:7096–7100

Navia MA, Fitzgerald PMD, McKeever BM, Leu C-T, Heimbach JC, Herber WK, Sigal IS, Darke PL, Springer JP (1989): Three dimensional structure of aspartyl protease from human immunodeficiency virus HIV-1. *Nature (London)* 337:615–620

Nunberg JH, Schleif WA, Boots EJ, O'Brien JA, Quintero JC, Hoffman JM, Emini EA, Goldman ME (1991): Viral resistance to human immunodeficiency virus type 1-specific pyridinone reverse transcriptase inhibitors. *J Virol* 65:4887–4892

Oberg B (1989): Antiviral effects of phosphonoformate (PFA, Foscarnet sodium). *Pharmacol Ther* 40:213–285

O'Brien JA, Ostovic D, Schorn TW, Smith SJ, Ruffing TL, Siegl PKS, Goldman ME (1992): A rapid bioassay for the determination of non-nucleoside HIV-1 reverse transcriptase inhibitor plasma levels. *Life Sci* 52:243–249

Pauwels R, Andries K, Desmyter J, Schols D, Kukla MJ, Breslin HJ, Raeymaeckers A, Van Gelder J, Woestenborghs R, Heykants J, Schellenkens K, Janssen MAC, De Clercq E, Janssen PAJ (1990): Potent and selective inhibition of HIV-1 replication *in vitro* by a novel series of TIBO derivatives. *Nature (London)* 343:470–474

Pettibone DJ, Clineschmidt BV, Anderson PS, Freidinger RM, Lundell GF, Koupal LR, Schwartz CD, Williamson JM, Goetz MA, Hensens OD, Liesch JM, Springer JP (1989): A structurally unique, potent and selective oxytocin antagonist derived from *Streptomyces silvensis. Endocrinology (Baltimore)* 125:217–222

Saari WS, Hoffman JM, Wai JS, Fisher TE, Rooney CS, Smith AM, Thomas CM, Goldman ME, O'Brien JA, Nunberg JH, Quintero JC, Schleif WA, Emini EA, Stern AM, Anderson PA (1991): 2-Pyridinone derivatives: A new class of nonnucleoside, HIV-1-specific reverse transcriptase inhibitors. *J Med Chem* 34:2922–2925

Saari WS, Wai JS, Fisher TE, Thomas CM, Hoffman JM, Rooney CS, Smith AM, Jones JH, Bamberger DL, Goldman ME, O'Brien JA, Nunberg JH, Quintero JC, Schleif WA, Emini EA, Anderson PS (1992): Synthesis and evaluation of 2-pyridinone derivatives as HIV-1-specific reverse transcriptase inhibitors. 2. Analogues of 3-aminopyridin-2(1*H*)-one. *J Med Chem* 35:3792–3802

Spadari S, Sala F, Pedrali-Noy G (1982): Aphidicolin: A specific inhibitor of nuclear DNA replication in eukaryotes. *Trends Biochem Sci* 7:29–32

White EL, Buckheit RW Jr, Ross LJ, Germany JM, Andries K, Pauwels R, Janssen PAJ, Shannon WM, Chirigos MA (1991): A TIBO derivative, R82913, is a potent inhibitor of HIV-1 reverse transcriptase with heteropolymeric templates. *Antiviral Res* 16:257–266

Wieland T, Gotzendorfer C, Zanotti G, Vaisius AC (1981): The effect of the chemical nature of the side chains of amatoxins in the inhibition of eukaryotic RNA polymerase B. *Eur J Biochem* 117:161–164

6

Discovery and Development of the HIV Proteinase Inhibitor Ro 31-8959

NOEL A. ROBERTS AND SALLY REDSHAW

Characterization of Human Immunodeficiency Virus

Since 1984, when the human immunodeficiency virus (HIV) became generally available to research laboratories, the full power of modern molecular biology and genetic analysis has been focused on the 9749 nucleotides which make up the viral genome. The first descriptions of HIV provided evidence for reverse transcriptase activity as well as circumstantial evidence for the existence of a viral proteinase. In 1985, molecular cloning and sequence determination revealed putative *gag* and *pol* open reading frames. In the known retroviruses, these open reading frames are first translated into the *gag* and *gag-pol* fusion polyproteins, which are subsequently processed to mature proteins, at least in part by a virally encoded proteinase. On the basis of the nucleotide sequence, it was suggested (Ratner et al., 1985) that the *pol* open reading frame encoded a proteinase analogous to those of other retroviral proteinases.

It was subsequently recognized (Toh et al., 1985) that the highly conserved Asp-Thr-Gly triad in the putative retroviral proteinase sequences was reminiscent of the catalytic center of aspartic proteinases, and it was suggested that the viral enzymes might belong to this family.

Despite this sequence homology, there appeared to be significant differences between the proposed viral enzymes and the classical aspartic proteinases. All of these latter proteinases comprise more than 200 amino acid residues and consist of two homologous domains. The key triad, Asp-Thr-Gly, occurs in each domain, and the two aspartic acids must come together to form the catalytic center. In contrast, the retroviral proteinases contain only around 100 amino acid residues and

The Search for Antiviral Drugs
Julian Adams and Vincent J. Merluzzi, Editors
© 1993 Birkhäuser Boston

a single Asp-Thr-Gly motif. These observations subsequently led to the proposal (Pearl and Taylor, 1987) that the retroviral proteinase must function as a homodimer, with each monomer contributing one aspartic acid residue to the active site.

Before the mechanism of action of the proteinase had been resolved, workers at Roche demonstrated that the proteinase in HIV *pol* was essential for the processing of the *gag* polyprotein (Kramer et al., 1986). Recombinant HIV *gag-pol* could be expressed in yeast, and processing of the *gag* polyprotein was demonstrated. However, when a frame-shift mutation was made in the proteinase region of *pol*, *gag* processing was lost. Thus, at least in yeast cells, the function of HIV proteinase could not be replaced by cellular proteinases and this enzyme was identified as a potential target for HIV therapy. Indeed, subsequent work at Roche showed the proteinase to be responsible for all the cleavages involved in the maturation of both the *gag* and *pol* gene products (Le Grice et al., 1988).

Support for the hypothesis that the proteinase is indeed essential for the viral life cycle came from site-directed mutagenesis experiments in which its active site aspartic acid was mutated (Kohl et al., 1988). Not only did these mutations abolish proteolytic activity, but when the mutant gene was reincorporated into the proviral DNA and used to transfect human colon carcinoma cells, no *gag* processing was observed and the resulting viral particles were noninfectious. Compelling evidence was thus accruing throughout the 1980s that HIV does indeed encode an aspartic proteinase and that this proteinase is mandatory for viral replication.

Development of the Roche AIDS Research Program

The Early Days (1985–1986)

Roche's initial commitment to AIDS research was the result of individual interests in basic research, diagnostics, new therapies, and the extended usage of existing therapies. The diagnostics groups in Basel, Switzerland, and Nutley, NJ, recognized the potential for improved clinical diagnostic systems for AIDS, and the Clinical Research and Development Departments began to investigate the use of Roferon A® as a treatment for Kaposi's sarcoma. The potential benefit of IL-2 in AIDS therapy was also being explored. Researchers from the Department of Molecular Genetics and Molecular Oncology in Nutley, in collaboration with the National Cancer Institute (NCI), commenced the studies on *gag* processing in yeast just described. In late 1985, Roche's Antiviral Chemotherapy Group in Welwyn, Hertfordshire, England, initiated a program to synthesize novel inhibitors of HIV reverse transcriptase (RT) (Martin et al., 1990; Wong-Kai-In et al., 1991). In support of this program, a collaboration was established in early 1986 with St. Mary's Hospital Medical School, London, to set up an antiviral testing facility.

Planning for Success (1986)

In May 1986, the current status of the AIDS pandemic was reviewed in Nutley with a call for a corporate commitment to AIDS research and the formation of an AIDS Task Force. A coordinated international program for the discovery of new therapeutic agents was envisaged, and the start of a Tat-inhibitor project in Nutley was proposed (see Chapter 7 of this volume). HIV proteinase had yet to emerge as a viable therapeutic target. In early June, a meeting of the International Antiviral Contact Group was held in the United Kingdom. Welwyn scientists reported their progress with RT inhibitors and also reviewed other possible molecular targets in the life cycle of HIV. These included its proteinase. This meeting did not initiate a program to design HIV proteinase inhibitors, but the stage had been set.

The importance of AIDS as a therapeutic target was again stressed at an International Anti-infectives Strategy Meeting held in Nutley at the end of June. Discussions between scientists and research managers from both Welwyn and Nutley were held throughout July and August: it was agreed that the most effective attack on AIDS could be made through close collaboration between the centers. Welwyn would continue to work on RT inhibitors with molecular biological support from Basel and would also assume responsibility for the synthesis and screening of proteinase inhibitors. Nutley would continue with their Tat-inhibitor program (Hsu et al., 1991) and would also become involved with the molecular biology of HIV proteinase.

The details of these proposals for a concerted effort against AIDS were presented to Hoffman La Roche Senior Research Management in October 1986. The plan was approved and a management leader (from Welwyn) and a scientific leader (from Nutley) were appointed for the joint program. A truly international collaboration had been established.

HIV Proteinase

Design for Discovery

In 1972, a basic research program had been initiated in Welwyn to explore the possibility of designing inhibitors of proteinases for which the protein crystal structure, the substrate specificity, or both, were known. These early studies (Hassall et al., 1979), which involved close collaboration between peptide chemists, biochemists, and molecular modeling experts, used the serine proteinase porcine pancreatic elastase as a model system.

The lessons learned from this pioneering work were later applied to a number of real therapeutic targets. Inhibitors of human leukocyte elastase were prepared (Kennedy et al., 1987; Hassall et al., 1985) as potential therapy for pulmonary emphysema, and prototype inhibitors of the aspartic proteinases renin and cathepsin D and the cysteine proteinase cathepsin B were also designed and synthesized. A major program on the inhibition of the metalloproteinase angiotensin-converting enzyme (ACE) culminated in the very potent and selective inhibitor, cilazapril (Attwood et al., 1986) now launched as Inhibace® or Vascace®. A current program on the inhibition of matrix metalloproteinases (Johnson et al., 1987) includes collagenase as a target enzyme and offers promise of a genuine treatment for rheumatoid arthritis and osteoarthritis.

Within the Research Division in Welwyn there was thus considerable expertise in the strategy and methodology of proteinase inhibition as well as a healthy awareness of the possible pitfalls.

Proteinase as a Target

Information about the structure and function of HIV proteinase was far from complete when our inhibitor program began in 1986. The proteinase had been provisionally classified as an aspartic proteinase on the basis of an Asp-Thr-Gly motif in its sequence, but exactly how an active catalytic site could be formed from only one such triad was as yet unclear. Two putative cleavage sites in the *gag* protein had been identified (Sanchez-Pescador et al., 1985), and further cleavage sequences were later determined in the *pol* protein (Lightfoote et al., 1986; Veronese et al., 1986). The enzyme had been shown to process the *gag* gene product, but definite proof that the enzyme was essential for viral growth was still lacking.

We were optimistic that we could design inhibitors of this enzyme using the transition state mimetic concept which had proved to be so successful for inhibitors of other aspartic proteinases, notably renin. We were also confident that we would be able to develop appropriate test systems in which to evaluate our inhibitors. One of the attractive features about HIV proteinase as a therapeutic target was felt to be its ability to effect cleavages between Phe-Pro and Tyr-Pro. Endopeptidase cleavage N-terminal to proline is a very rare event in mammalian biochemistry, so inhibitors based on these cleavage sequences would be expected to be highly selective for the viral enzyme and should not significantly inhibit the mechanistically similar human proteinases renin, pepsin, gastricsin, cathepsin D, or cathepsin E. Pepsin, gastricsin, and cathepsin E are all present in high concentration in the gastrointestinal tract: binding to these enzymes might thus reduce the chances of oral absorption of an inhibitor, and it was already known from studies on renin inhibitors that peptide mimetics similar to those proposed as inhibitors of HIV proteinase tended to show poor oral bioavailability.

The promise of selectivity offered by inhibitors based on a Phe-Pro cleavage sequence was thus very attractive to us. However, this putative cleavage site for HIV proteinase had not been unequivocally proven at this time, merely inferred from sequence data and by comparison with other retroviral proteinases. One of our top priorities was therefore to confirm this cleavage sequence either by site-specific mutagenesis of *gag* or by demonstrating this cleavage in synthetic peptide substrates.

Time Scales Defined

From the very outset, everyone involved in the proteinase inhibitor program, both managers and scientists alike, felt a keen sense of urgency and excitement, and the schedule set for the program was expeditious. Beginning in November 1986, our critical issues and time scales were as follows:

1. Cloning and expression of the proteinase; synthesis of peptides of no more than 10 residues spanning the putative cleavage sites and, using semipurified enzyme, demonstration of the cleavage of Phe-Pro or Tyr-Pro
   Projected completion: mid-1987
2. Purification of the proteinase and development of a rapid, preferably colorimetric assay; synthesis and evaluation of inhibitors, achieving potent ($K_i = 10^{-7}$–10^{-9} M) and selective [K_i(human aspartic proteinases)/K_i(HIV-PR) > 100] inhibitors
   Projected completion: mid-1988
3. Screening of inhibitors for anti-HIV activity and demonstration of inhibition of protein processing
   Projected completion: end of 1988
4. Identification of inhibitors with suitable pharmacokinetic profile and selection of candidate for clinical development
   Projected completion: end of 1990

We had thus allowed ourselves 4 years from inception to identification of a clinical candidate. We never fell behind this schedule, and an inhibitor (Ro 31–8959) was proposed for development in September 1989, just 35 months from the start of the program.

Putting the Plan into Action

Late in 1986, work began in a spirit of mutual confidence and optimism. The enzyme was cloned and expressed in the surety that there would be an assay to detect it, and assay systems were developed in the confidence that Phe-Pro would indeed be cleaved and that the enzyme would be made available. Synthesis of transition state mimetics based on Phe-Pro also began on the assumption not only that enzyme and assay would become available but also that Phe-Pro was indeed an HIV proteinase cleavage site. We were all delighted (and somewhat relieved!) when, just 10 months later, our hypotheses and suppositions were confirmed and we first detected the inhibition of HIV proteinase with one of our novel compounds. This period was perhaps more tense than any subsequent phase of the project since failure then in any of our objectives could have ended the whole program.

Cloning and Expression of HIV Proteinase

While the Roche molecular biologists at Nutley had set out specifically to clone, express, and purify HIV proteinase (Graves et al., 1988), it became apparent that some of the *pol* constructs available in Basel might also be used to provide a source of proteinase (Mous et al., 1988). Both centers were soon supplying Welwyn with plasmids which would potentially express HIV proteinase. A number of these gave a protein of the correct molecular weight as determined by Western blotting analysis, but the first plasmid to generate proteinase activity detectable in our colorimetric assay was pPTΔN.

This plasmid was subsequently used as the source of both crude and purified enzyme for our inhibition studies. The Basel clones (Le Grice et al., 1988) were used in many of the studies into the mode of action of the enzyme.

Enzyme Assays

The design of a rapid, easily quantified, and preferably colorimetric assay for an enzyme which has yet to be fully characterized in terms of pH optimum or substrate specificity imposes severe limitations and forces major assumptions. A discontinuous assay seemed preferable, as the pH and other conditions for the color reaction could then be optimized independently of the conditions best suited to the enzyme function. Our presumptions about the enzyme's substrate specificities meant that the ability to detect cleavage of a Phe-Pro bond was clearly our best option. The detection of newly formed N-terminal proline seemed especially attractive. Since N-terminal proline is not a common feature in standard peptides and proteins, such an assay should not only be applicable to a range of potential substrates including proteins but should also be viable using crude enzyme preparations since any extraneous protein would not create high blank values.

The quantitative reaction of isatin with free proline to give a blue color had been known since the end of the last century, but the reaction with peptides containing N-terminal proline required harsher conditions (Grassmann and von Arnim, 1935) and had previously only been applied qualitatively. We were able to achieve a quantitative color reaction using relatively mild conditions by identifying (largely through trial and error!) a suitable acid catalyst (Broadhurst et al., 1991). Protected hexapeptide (1) based on a conserved P_4–P'_2 sequence around the (putative) Phe(Tyr)-Pro cleavage sites in *gag* and *pol* polyproteins was chosen as our original substrate.

Succinyl.Ser.Leu.Asn.Tyr.Pro.Ile.NH^{i}Bu

(1)

Protection at both N and C termini was designed to prevent adventitious cleavage of our substrate when used with crude enzyme preparations, and the N-terminal succinyl group was chosen to enhance water solubility. This "best guess" compound was indeed a reasonable substrate for the viral proteinase (K_m = 1.42 mM), and was used for most of the initial enzyme characterisation as well as our early inhibitor studies. A protected heptapeptide (2) was subsequently found (Broadhurst et al., 1991) to be a better substrate (K_m = 0.79 mM), but even using this improved substrate, the assay is affected by mutual depletion with inhibitors of K_i 1.0 nM.

Succinyl.Val.Ser.Gln.Asn.Phe.Pro.Ile.NH^{i}Bu

(2)

In addition to this high-throughput assay, two further assays were set up to measure processing of the natural substrates of the proteinase. These were a mixed lysate assay of *gag* protein processing, and a whole cell assay employing expression of a *gag-pol* protein in a baculovirus system (Overton et al., 1990).

Chemistry and SARs

Working on the as yet unproven assumption that the HIV proteinase was indeed an aspartic proteinase with the unusual ability to cleave peptides between an aromatic amino acid and proline, we set about preparing simple peptides incorporating different transition state mimetics based on the putative Asn.Phe-Pro cleavage sequence in the *pol* polyprotein. Some of these early tripeptide mimetics showed very promising activity (Table 1, compounds 3–6). We also considered using

TABLE 1. Comparative activities of various transition state mimetics incorporated into a simple tripeptide analogue

Compound	Structure	Mimetic type	IC$_{50}$ (μM)
3		Reduced amide	50
4		Ketomethylamine	0.87
5		Hydroxyethylamine	0.3
6		Hydroxyethylamine	0.14

the hydroxyethylene (7) or phosphinate (8) transition state mimetics. Neither of these readily accommodates the imino acid moiety which we predicted would confer selectivity for the viral proteinase. Compounds of the hydroxyethylene class, although potent inhibitors of HIV proteinase, were also surrounded by a veritable minefield of renin-inhibitor patents. We found that inhibitors containing a phosphinate moiety, although very potent, were not selective for the viral enzyme. More importantly, the phosphinate derivatives also showed very low antiviral activity, probably because of poor cell penetration.

We were persuaded by these findings that our best opportunity for success lay with the hydroxyethylamine transition state mimetic. Although our prototype inhibitor (6) showed very encouraging enzyme inhibitory potency, we felt that for clinical assessment, especially with a new class of therapeutic agent, a more potent compound would be required. At this time, there was no X-ray crystallographic data available either for the native enzyme or for enzyme–inhibitor complexes (the first of these was to appear in 1989: Navia et al., 1989), so we set out to modify our lead structure systematically by modifying each residue in turn (Roberts et al., 1990). In all, we prepared some 200 compounds, a selection of which are shown in Table 2. We found that an aromatic

TABLE 2. Effect of modification of individual residues on inhibitory potency[a]

Compound	Stereochemistry at $-CH(OH)-$	P_3	P_2	P_1		P_1'	P_2'	IC_{50} (nM)
6	R	Z.	Asn.	Phe. Ψ [CH(OH)CH$_2$N]		Pro.	O^{t}Bu	140
9	R	BOC.	Asn.	Phe. Ψ [CH(OH)CH$_2$N]		Pro.	O^{t}Bu	8000
10	R	Ts.	Asn.	Phe. Ψ [CH(OH)CH$_2$N]		Pro.	O^{t}Bu	2200
11	R	PhCO.	Asn.	Phe. Ψ [CH(OH)CH$_2$N]		Pro.	O^{t}Bu	500
12	R	Quin.	Asn.	Phe. Ψ [CH(OH)CH$_2$N]		Pro.	O^{t}Bu	23
13	R	Z.	Ala.	Phe. Ψ [CH(OH)CH$_2$N]		Pro.	O^{t}Bu	2000
14	R	Z.	Gln.	Phe. Ψ [CH(OH)CH$_2$N]		Pro.	O^{t}Bu	1200
15	R	Z.	Phe.	Phe. Ψ [CH(OH)CH$_2$N]		Pro.	NH^{t}Bu	420
16	R	Z.	Asn.	Cha. Ψ [CH(OH)CH$_2$N]		Pro.	O^{t}Bu	1200
17	R	Z.	Asn.	Hph. Ψ [CH(OH)CH$_2$N]		Pro.	O^{t}Bu	>1000
18	R	Z.	Asn.	Omt. Ψ [CH(OH)CH$_2$N]		Pro.	O^{t}Bu	140
19	R	Z.	Asn.	Phe. Ψ [CH(OH)CH$_2$N]		Phe.	O^{t}Bu	>1000
20	R	Z.	Asn.	Phe. Ψ [CH(OH)CH$_2$N]		^{HO}Pro.	O^{t}Bu	100
21	R	Z.	Asn.	Phe. Ψ [CH(OH)CH$_2$N]		PIP.	NH^{t}Bu	18
22	R	Z.	Asn.	Phe. Ψ [CH(OH)CH$_2$N]		DIQ.	NH^{t}Bu	<2.7
23	S	Z.	Asn.	Phe. Ψ [CH(OH)CH$_2$N]		DIQ.	NH^{t}Bu	>1000
24	R	Z.	Asn.	Phe. Ψ [CH(OH)CH$_2$N]		Pro.	NHPh	2500
25	R	Z.	Asn.	Phe. Ψ [CH(OH)CH$_2$N]		Pro.	N(Me)^{t}Bu	>1000
26	R	Z.	Asn.	Phe. Ψ [CH(OH)CH$_2$N]		Pro.	NHMe	670
27	R	Z.	Asn.	Phe. Ψ [CH(OH)CH$_2$N]		Pro.	NH^{t}Bu	70

[a]Abbreviations: Z, benzyloxycarbonyl; BOC, tert-butyloxycarbonyl; Ts, *p*-toluenesulfonyl; Quin, quinoline-2-carbonyl; Cha, cyclohexylalanine; Hph, homophenylalanine; Omt, *O*-methyltyrosine; PIP, piperidin-2(*S*)-ylcarbonyl; DIQ, [(4a*S*, 8a*S*)-decahydroisoquinolin-3(*S*)-yl]carbonyl.

group was strongly preferred at the S_3 subsite. At the S_2 subsite, no improvement over asparaginyl was found; similarly at P_1, no significant improvement was found over the benzyl side chain of phenylalanine. Early modeling studies suggested that the enzyme should be able to accommodate amino acids other than proline at P'_1 [this was later confirmed by X-ray crystallographic studies on the native enzyme in which the perfect twofold rotational (C_2) symmetry means that the S_1 and S'_1 subsites are indistinguishable]. Dramatic changes in potency were indeed achieved by varying the imino acid at P'_1, particularly effective replacements for prolyl being piperidin-2(S)-ylcarbonyl (compound 21) and [(4aS,8aS)-decahydroisoquinolin-3(S)-ylcarbonyl] (compound 22). At the carboxyl terminus, the *tert*-butyl ester could be replaced by a *tert*-butyl amide without loss of potency, but again no improvement was found.

Interestingly, in this series of compounds, the stereochemical requirement at the carbon atom bearing the hydroxyl group is opposite to that in the hydroxyethylene-containing inhibitors, or indeed in hydroxy-ethylamine-containing compounds which are further extended toward the carboxyl terminus (Rich et al., 1990, 1991; Krohn et al., 1991). The small differences in potency which we observed in our prototype inhibitors (compounds 5 and 6) become much more pronounced in the more active inhibitors (compounds 22 and 23). We were delighted to discover that in this series of compounds the improvements in individual residues were additive, so that by combining preferred side chains we were able to achieve highly potent inhibitors.

A patent was filed in June 1988 covering these inhibitors generically (Handa et al., 1988) and our preferred compound (28; Ro 31–8959) was specifically claimed in a patent filed in December 1989 (Martin and Redshaw, 1989). Our synthetic route to Ro 31–8959 starting from phenylalanine is outlined in Scheme 1.

Antiviral Assays

By mid-1987, the laboratory at St. Mary's Hospital was routinely testing our RT inhibitors for antiviral activity, using syncytial count as an end point. An ELISA (enzyme-linked immunosorbent assay) method for detection of p24 antigen had also been developed to provide an alternative end-point measurement. Our first proteinase inhibitors were synthesized in the spring of 1987, about 6 months before the enzyme-assay was to become available. The temptation was too great! Several of the compounds were tested for antiviral activity before we had evidence that they were inhibitors of the proteinase. Although weak ($ED_{50} =$ 25–100 μM), antiviral activity was demonstrated and the chemical program continued with increased optimism.

As soon as our own P3 containment laboratory was commissioned, we were able to begin in-house testing of our proteinase inhibitors.

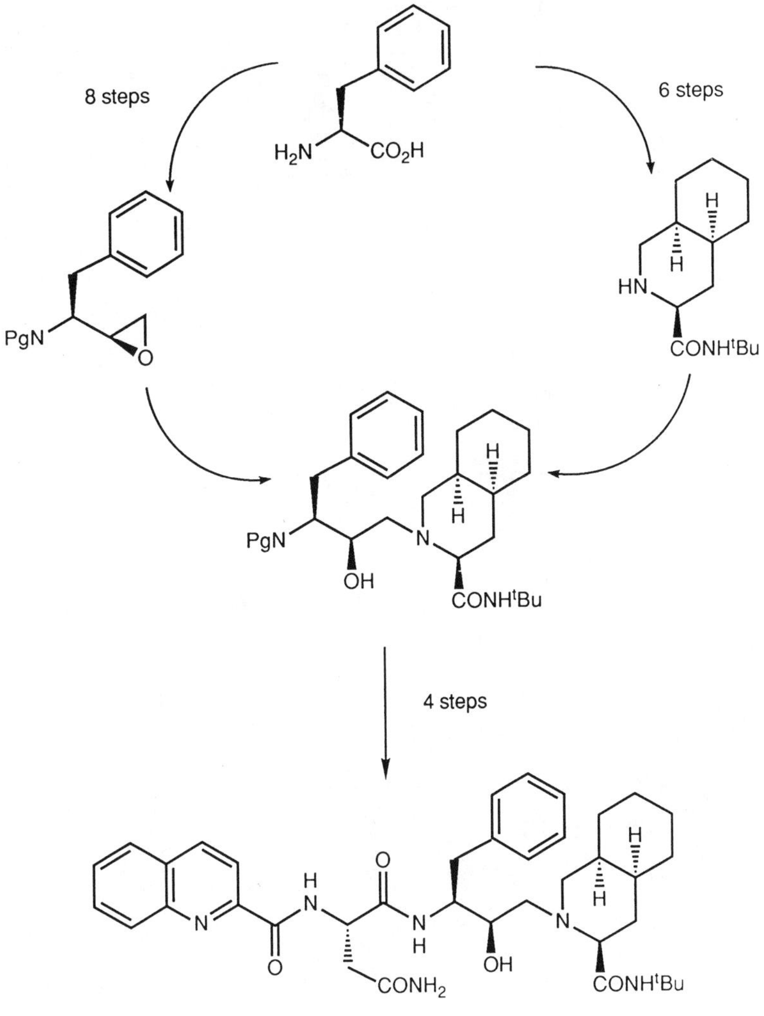

(28)

Pg = Protecting Group

SCHEME 1. Synthetic route to Ro 31-8959.

Using a variety of cell lines and viral strains, we were able to demonstrate antiviral activity for our compounds utilizing both syncytial counts and p24 antigen levels as end points (Table 3). Happily, our own results and those of external collaborators showed that the antiviral activities of our compounds closely paralleled the IC_{50} values against the isolated enzyme (Table 4).

TABLE 3. Antiviral activities of Ro 31–8959 in various virus–host cell systems

Test center (virologist)	Cell line	Virus strain	Assay method	IC$_{50}$ (nM)	
				Ro 31–8959	Standard
Roche Welwyn (Craig)	JM	HIV-1 GB8	Syncytia	3.1 ($n = 10$)[a]	AZT > 1000
	JM	JM/HIV-1 GB8 (infectious center assay)	Syncytia	0.66 ($n = 4$)	AZT > 1000
	CEM	HIV-1 GB8	Syncytia p24 antigen	7.8 ($n = 18$) 4.0 ($n = 19$)	AZT = 12.9 ($n = 19$) AZT = 5.9 ($n = 8$)
	SUP-T1	HIV-1 GB8	Syncytia	2.0	–
	C8166	HIV-1 RF	p24 antigen	8.0 ($n = 2$)	4.2
	AA2	HIV-1 RF	p24 antigen	(IC$_{90}$ = 1.6)	(AZT IC$_{90}$ = 2.1)
	JM	HIV-1 MN	Syncytia	3.7	–
	MT-4	HIV-1 RF	Plaques	20.5	AZT = 12.0
Roche Basel (Mous)	MT-4	HIV-1 III$_B$	CPE (MTT)	12 ($n = 3$)	AZT = 4.8 ($n = 3$)
	Jurkat	HIV-2 Ben	RT activity	0.45 ($n = 2$)	AZT = 6
St. Mary's Hospital, London (Kinchington[b])	C8166	HIV-1 RF	p24 antigen	1.7 ($n = 4$)	AZT = 1
	C8166	HIV-2 ROD	p24 antigen	4	–
	MT-2	HIV-1 A018 (AZT^R CI)[c]	p24 antigen	1–10	AZT = 1,000–10,000
MRC Mill Hill (Tyms)	JM	HIV-1 GB8	p24 antigen	0.7	ddC = 150
	JM	HIV-1 GB8	Syncytia	2.8 ($n = 2$)	ddC = 440

[a]Number of experiments ($n = 1$ unless otherwise indicated).
[b]Present address: St. Bartholomew's Hospital, London.
[c]AZT^R = AZT resistant; CI = clinical isolate.

TABLE 4. Proteinase inhibitory activity and antiviral activity of selected compounds

Compound	Proteinase IC_{50} (nM)	Antiviral IC_{50} (nM)
5	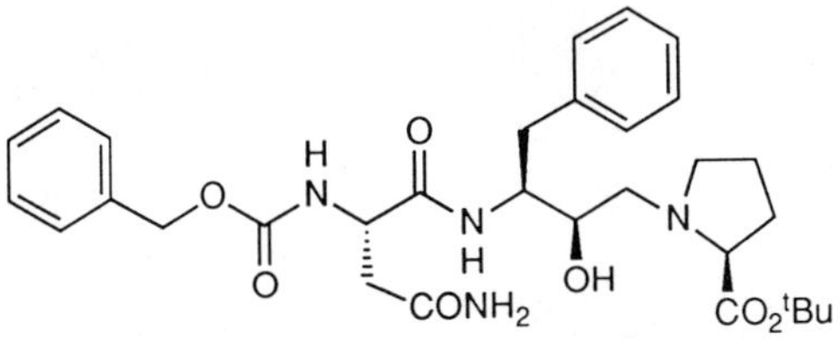 140	300
29	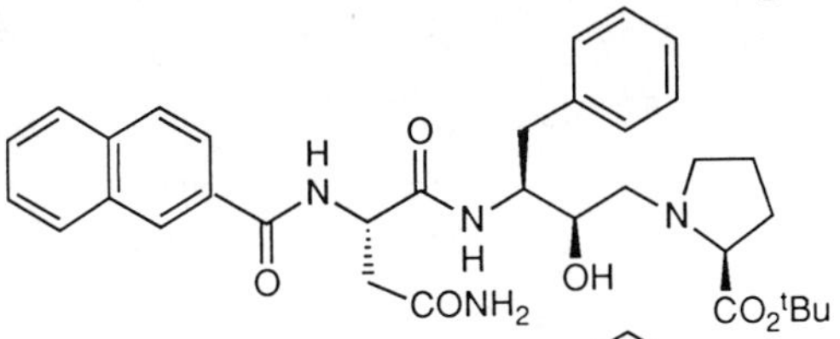 52	130
30	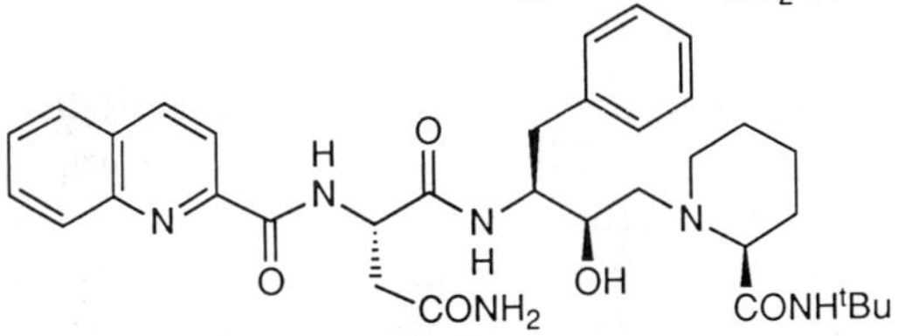 6.1	17
28	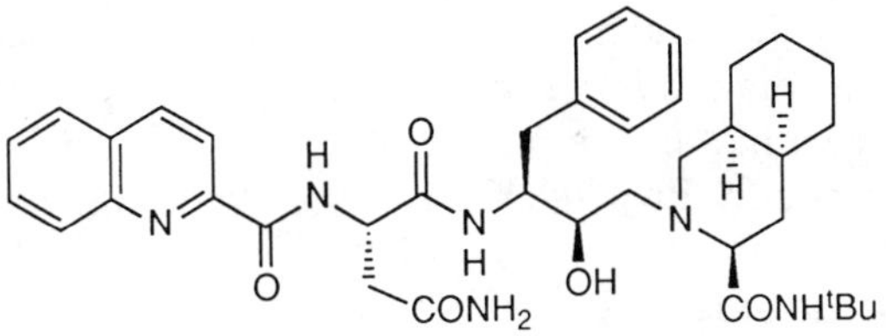 <0.4	2

Focusing In

Although the strategy of basing inhibitor design on Phe-Pro mimetics had promised selective proteinase inhibitors, we felt it was important to confirm this by testing our compounds for inhibitory selectivity against the human aspartic proteinases. Assays for inhibition of a number of mammalian aspartic proteinases, including human renin, were available in Roche, Basel, and our compounds were shown to be inactive in these tests. However, a much more comprehensive "set" of human aspartic proteinases was held by Dr. John Kay at the University of Cardiff (Wales). He was pleased to collaborate with us, and very rapidly assured us that most of our lead compounds, including Ro 31–8959, were indeed highly selective inhibitors of HIV proteinase (Table 5).

TABLE 5. Proteinase selectivity of Ro 31–8959

Proteinase	Class	IC_{50} (nM)
HIV-1	Aspartic	$<0.37^a$ ($K_i = 0.12$)
HIV-2	Aspartic	$<0.80^a$ ($K_i = 0.10$)
Human renin	Aspartic	$>10{,}000$
Human pepsin	Aspartic	$>10{,}000$
Human gastricsin	Aspartic	$>10{,}000$
Human cathepsin D	Aspartic	$>10{,}000$
Human cathepsin E	Aspartic	$>10{,}000$
Human leukocyte elastase	Serine	$>10{,}000$
Bovine cathepsin B	Cysteine	$>10{,}000$
Human synovial fibroblast collagenase	Metallo	$>10{,}000$
Prolidase	Dipeptidase	$>10{,}000$

^{a}Results limited by mutual depletion.

From the outset of our work on HIV, we had been acutely aware of the lack of a suitable animal model. This situation had not improved by the time we were trying to distinguish between several very potent inhibitors in order to recommend a compound for development. Unfortunately, at least from this viewpoint, our inhibitors had not shown significant activity against the murine retroviruses, which meant that we were unable to use this model. The particularly poor pharmacokinetic profile of our compounds in the mouse also made the SCID-hu model inappropriate. Although Ro 31–8959 is equipotent against HIV and simian immunodeficiency virus (SIV) (Martin et al., 1991), the effect of drugs on the pathogenesis of SIV in the monkey had not been sufficiently well characterized at that time to provide a validated model.

These factors meant that defining a target profile for a development candidate was problematic and had to be carried out in the absence of any real efficacy data. Although we had firm data on the enzyme inhibitory activity and antiviral potency of our compounds, we had no real basis for predicting how these values would relate to the clinical situation. Estimates of how many multiples of the IC_{50} (IC_{90}?) would need to be achieved in plasma (target tissue or cells?) proved controversial.

Our ultimate goal had always been an orally active antiviral agent: important determinants here were felt to be selectivity for the viral enzyme as well as the physiochemical properties of individual compounds.

Broadly speaking, oral absorption is promoted by good water solubility accompanied by a high partition coefficient—an unlikely specification for highly peptidic compounds (e.g., renin inhibitors). We were therefore concerned that one of the early assessments of our compounds should be a pharmacokinetic evaluation in animals. Because such a screen is necessarily rather labor intensive, we set fairly stringent criteria on enzyme inhibitory potency and antiviral activity before compounds would be submitted for pharmacokinetic analysis. These early studies allowed us to select Ro 31–8959 for further evaluation since a clinically realistic dose of 10 mg/kg p.o. gave, in the rat, plasma levels which exceeded the antiviral IC_{90} for about 6 hours after dosing.

The Development Compound: Recent Data

When Ro 31–8959 was proposed for development we still had much to discover about its biological properties and also about its chemical synthesis. Up to that time we had prepared only 10 g of material, and indeed most of the critical in vitro studies had been achieved with about 10 mg of compound! The antiviral data from three centers was limited to tests using laboratory strains of virus and cell lines, and in vivo experiments were restricted to pharmacokinetic studies. After considering various salt forms, we selected the mesylate for further galenic studies on the basis of its high crystallinity and considerably enhanced water solubility.

Despite the relatively limited data, Ro 31–8959 entered the development pipeline in late 1989 with the recommendation that a fast-track, high-risk approach be adopted, with every effort made to accelerate the development of the compound. In the weeks while the proposal documentation was being prepared and considered, work continued apace in anticipation of a favorable outcome!

The fate of the compound was now passing from the hands of the research scientists into those of an International Project Team. This team has full responsibility for the compound in its progress through preclinical development and formulation into toxicity studies in animals and then on to evaluation in healthy volunteers and clinical studies.

Drug development, particularly in a fast-track environment, is a major exercise in logistics. Sufficient material of approved specification must be made available to complete any toxicological or clinical studies, allowing even for a "worst-case" dose level scenario. For an aggressive development program, the rate of scaleup of synthesis must be expeditious. For Ro 31–8959, the total weight of compound synthesized has increased some 10,000-fold in 2 years (Figure 1). This achievement is the result of a considerable investment in time and expertise in both Welwyn and Basel.

In animal toxicity studies, the compound was shown to be sufficiently well tolerated to allow phase I clinical studies to be carried out in healthy

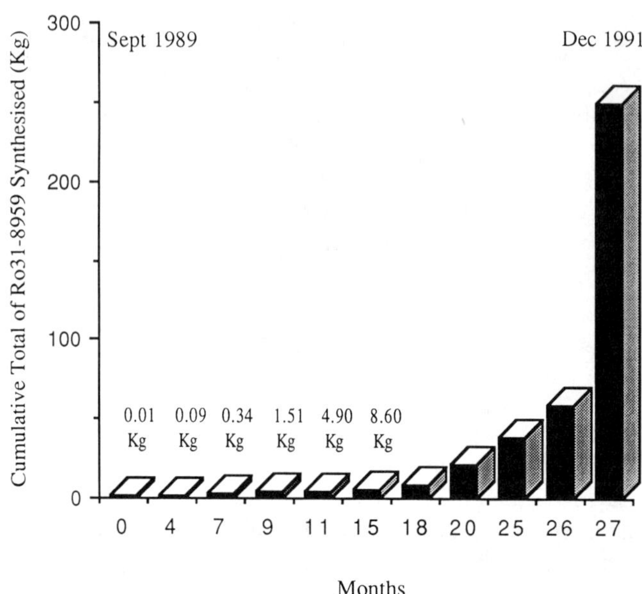

Figure 1. Cumulative total of Ro 31–8959 synthesized September 1989–December 1991.

volunteers. It was felt that this option would allow the most rapid progress since Roche has its own in-house volunteer panel and since biochemical and pharmacokinetic analysis of plasma samples can obviously be carried out more easily if the samples are noninfectious. These extensive volunteer studies are now complete. Data on tolerability and drug exposure levels obtained from this work have justified the initiation of international clinical trials in HIV-infected patients. The phase II studies have now begun, and the number of patients being treated with Ro 31–8959 is growing daily. The effect of the drug on viral and immunological markers will be determined when sufficient data have been accumulated and the double blind code is broken.

While work has been progressing on the direct route to clinical assessment, additional preclinical studies have also been undertaken to further our understanding of the biological properties of the compound. The structure of Ro 31–8959 bound to the active site of HIV proteinase (Figure 2) has been determined using X-ray crystallography (Krohn et al., 1991). As well as providing detailed information which might be used to develop second-generation inhibitors, the X-ray structure can also be used to study the positions at which mutations in the virus might affect inhibitor binding (J. Mous, A. Krohn, H. Jacobsen, unpublished results). Several laboratories have now assessed the antiviral activity of Ro 31–8959 in a range of different test systems (Holmes et al., 1991). It

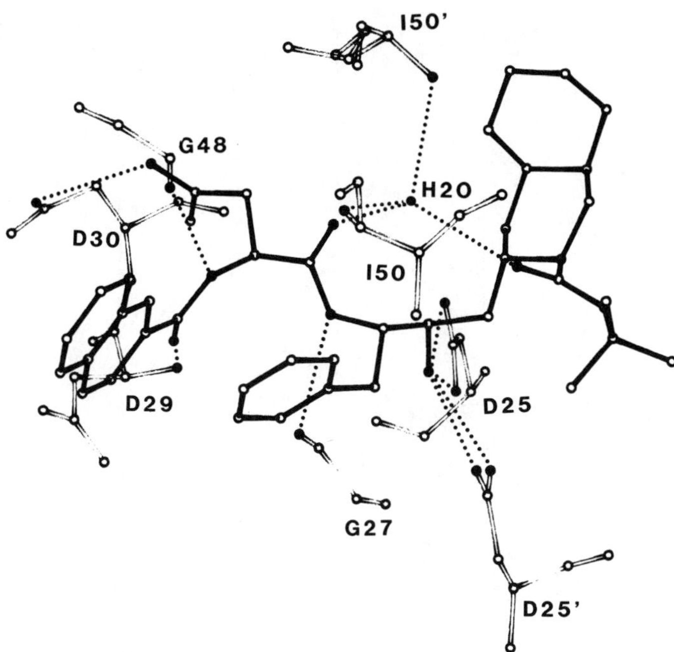

FIGURE 2. Ro 31–8959 (full bonds) shown bound to the active site of HIV proteinase (open bonds). Important H-bonding interactions between the enzyme and Ro 31–8959 are indicated by dotted lines.

is apparent that the compound has potent antiviral activity in both lymphoblastoid and macrophage cell lines as well as in primary cells, either acutely or chronically infected with laboratory strains of virus or with clinical isolates, including AZT-resistant strains and HIV-2 (Craig et al., 1991a,b; Johnson et al., 1992; Galpin et al., 1993). Typically, the IC_{50} value falls in the range 1–10 nM and the IC_{90} value in the range 5–100 nM (Table 6). When the compound is tested in combination with other antiviral agents, additive or synergistic action against HIV is seen (Roberts et al., 1992; Johnson et al., 1992). This is especially encouraging in the light of the generally held opinion that combination therapy may be the strategy of choice in the treatment of AIDS in order to limit both drug toxicity and the likelihood of drug resistance.

Studies have also continued both to characterize the mode of action of Ro 31–8959 and to use this specific inhibitor to clarify the role of HIV proteinase in the viral life cycle. Studies in which drug addition was delayed postinfection showed that Ro 31–8959 acts at a late stage in the virus life cycle, consistent with inhibition of virus maturation (Galpin et al., 1993). Electron microscopy studies of chronically infected lymphocytic and monocytic cell lines treated with Ro 31–8959 have given

TABLE 6. Antiviral and cytotoxicity data on Ro 31–8959 from an MRC multicenter blinded testing program (Holmes et al., 1991)

Center[b]	Cell line/ virus strain	End point	IC_{50} $(CC_{50})^a$ (nM)		Therapeutic index	
			Ro 31–8959	Standards	Ro 31–8959	Standards
1	C8166/HIV-1 RF	p24 antigen	3 (4.4×10^4)	AZT 3.7 $(>7.5 \times 10^5)$ ddC 190 $(>9.5 \times 10^5)$	1.5×10^4	AZT $>2 \times 10^5$ ddC $>5 \times 10^3$
2	C8166/ HIV-1 RF	Syncytia p24 antigen	<1.5 (7.5×10^4)	AZT 18.7 $(>1.9 \times 10^6)$ ddC 23.7 $(>2.4 \times 10^6)$	$>5 \times 10^4$	AZT $>10^5$ ddC $>10^5$
	JM/ HIV-1 GB8	Syncytia p24 antigen	<1.5 (7.5×10^4)	AZT 1.9×10^6 $(>3.7 \times 10^6)$ ddC <379 (2.4×10^3)	$>5 \times 10^4$	AZT >2 ddC >6
3	Molt 4/ HIV-1 NKR	Cytopathic Effect	$<1.5 \times 10^3$ (4.1×10^5)	AZT <37.4 (2.1×10^4) ddC <47.4 (474)	>280	AZT >550 ddC >10
4	MT4/ HIV-1 IIIB	Cell Viability	<0.75 (1.3×10^4)	AZT 1.1 (5.6×10^3) ddC 227 (4.5×10^4)	$>1.7 \times 10^4$	AZT 5×10^3 ddC 200
	MT4/ HIV-2$_{ROD}$	Cell Viability	<1.2 (1.8×10^4)	AZT 1.3 (4.9×10^3) ddC 275 (4.5×10^4)	$>1.5 \times 10^4$	AZT 3.6×10^3 ddC 170

[a]IC = inhibitory concentration; CC = cytotoxic concentration.
[b]MRC-appointed study centers:
1, St. Bartholomew's Hospital, London (formerly St. Mary's Hospital);
2, MRC Collaborative Centre, Mill Hill;
3, University of Cambridge;
4, Rega Institute for Medical Research.

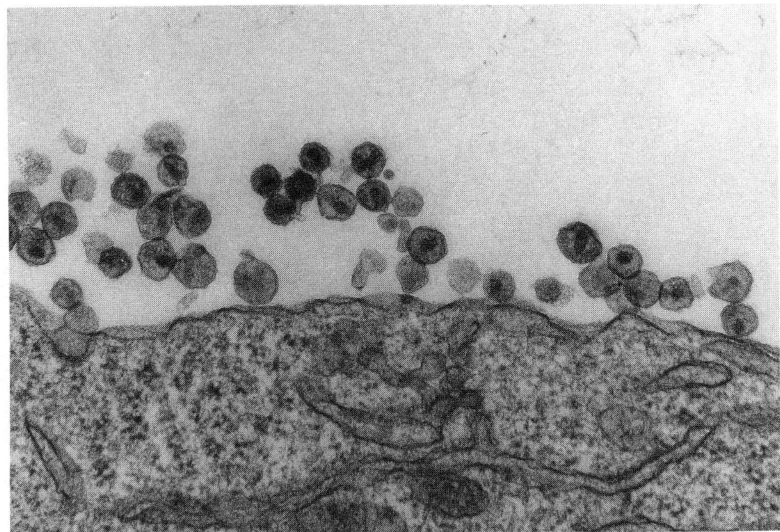

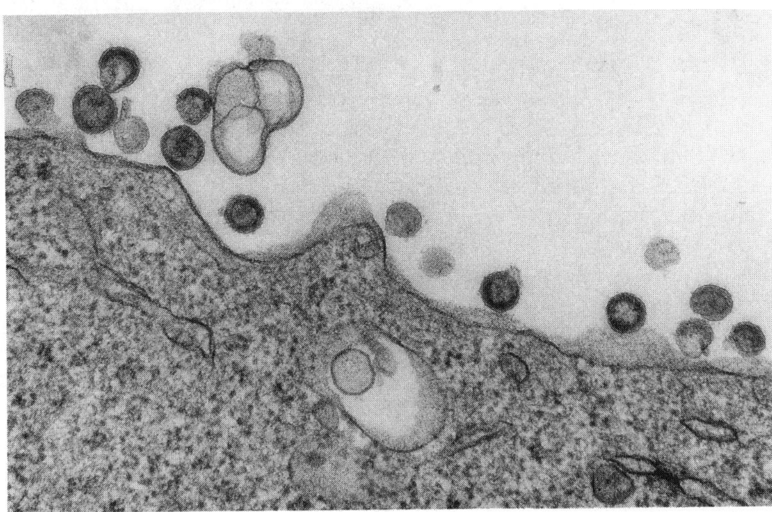

Figure 3(a) and (b). See next page for Figure 3(c) and caption.

clear visual evidence (Figure 3) that the compound arrests the maturation of virus particles, blocking conversion of the immature noninfectious "doughnut" form of the virions into infectious virus particles with a fully formed capsid (Craig et al., 1991a,b). The noninfectious nature of these immature virions produced in the presence of Ro 31–8959 has also been demonstrated directly by viral titer (Roberts et al., 1992). These virions do not regain infectivity when removed from the inhibitor. No evidence could be found of a role for the proteinase in the early stages of HIV infection of cells (Jacobsen et al., 1992).

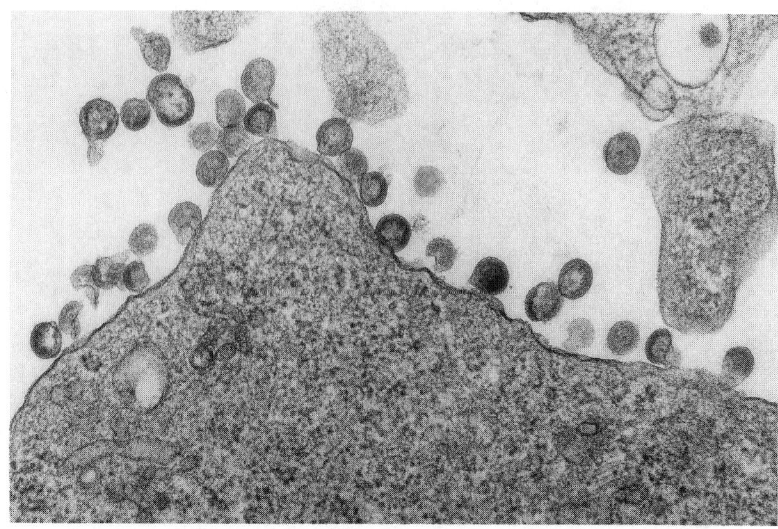

FIGURE 3(c). CEM cells chronically infected with HIV-1$_{\text{III}_\text{B}}$: (a) untreated—particles have a condensed core which is typical of mature virus; (b) treated for 24 hours with Ro 31–8959 (10 nM); (c) treated for 24 hours with Ro 31–8959 (100 nM)—particles have an electron dense outer layer which is typical of immature virus. Electron microscopy by D. Hockley and C. Grief. Magnification: (a)–(c), × 40,000. Reproduced with permission of Elsevier Science Publishers from Craig JC et al. (1991b): Antiviral properties of Ro 31–8959, an inhibitor of human immunodeficiency virus (HIV) proteinase. *Antivir Res* 16:295–305.

Thus, all the new data which have been accumulated since the decision was made to develop Ro 31–8959 has further supported that decision. Looking optimistically at the prospects for this compound, it could, at a dose very well tolerated by patients, markedly reduce the spread of virus in a manner largely independent of infected cell type. This may give the patient's immune system the opportunity to recover. The chronically infected cells will continue to produce viral antigen (in a noninfectious form), and this could further stimulate the patient's immune system to act against the infection. During treatment there should be a much reduced risk of infection to the patient's health-care workers and social contacts. Although there is still much work to be done, we remain very hopeful that Ro 31–8959 will make a positive contribution to the therapy of AIDS.

Acknowledgments

The major responsibility for the HIV proteinase program has been with Roche's Welwyn Research, but the successful and timely outcome of the work must be attributed to a substantial collaborative effort not only within Roche internationally but also with external academic groups and national AIDS programs. Adequate personal citation is not possible, but it has been the goodwill, inspiration, and hard work of the individual that has been the driving force of this program.

References

Attwood MR, Hassall CH, Krohn A, Lawton G, Redshaw S (1986): The design and synthesis of the angiotensin converting enzyme inhibitor cilazapril and related bicyclic compounds. *J Chem Soc [Perkin 1]* pp 1011–1019

Broadhurst AV, Roberts NA, Ritchie AJ, Handa BK, Kay C (1991): Assay of HIV-1 proteinase: A colorimetric method using small peptide substrates. *Anal Biochem* 193:280–286

Craig JC, Grief C, Mills JS, Hockley D, Duncan IB, Roberts NA (1991a): Effects of a specific inhibitor of HIV proteinase (Ro 31–8959) on virus maturation in a chronically infected promonocytic cell line (U1). *Antiviral Chem Chemother* 2:181–186

Craig JC, Duncan IB, Hockley D, Grief C, Roberts NA, Mills JS (1991b): Antiviral properties of Ro 31–8959, an inhibitor of human immunodeficiency virus (HIV) proteinase. *Antiviral Res* 16:295–305

Galpin S, Roberts NA, O'Connor T, Jeffries DJ, Kinchington D (1993): Antiviral properties of the HIV-proteinase inhibitor Ro 31-8959. *Antiviral Chem Chemother*, in press

Grassmann W, von Arnim K (1935): Über neue Farbreaktionen des Pyrrolidins und Prolins. *Justus Liebigs Ann Chem* 519:192–208

Graves MC, Lim JJ, Heimer EP, Kramer RA (1988): An 11-kDa form of human immunodeficiency virus protease expressed in *Escherichia coli* is sufficient for enzymatic activity. *Proc Natl Acad Sci USA* 85:2449–2453

Handa BK, Machin PJ, Martin JA, Redshaw S, Thomas GJ (1988): Amino acid derivatives. Eur Pat Appl. 0,346,847

Hassall CH, Johnson WH, Roberts NA (1979): Some novel inhibitors of porcine pancreatic elastase. *Bioorg Chem* 8:299–309

Hassall CH, Johnson WH, Kennedy AJ, Roberts NA (1985): A new class of inhibitors of human leucocyte elastase. *FEBS Lett* 183:201–205

Holmes HC, Mahmood N, Karpas A, Petrik J, Kinchington D, O'Connor T, Jeffries DJ, Desmyter J, De Clerq E, Pauwels R, Hay A (1991): Screening of compounds for activity against HIV: A collaborative study. *Antiviral Chem Chemother* 2:287–293

Hsu M-C, Schutt AD, Holly M, Slice LW, Sherman MI, Richman DD, Potash MJ, Volsky DJ (1991): Inhibition of HIV replication in acute and chronic infections *in vitro* by a Tat antagonist. *Science* 254:1799–1802

Jacobsen H, Ahlborn-Laake L, Gugel R, Mous J (1992): Active viral protease is not essential in early steps of HIV infection. *J Virol,* 66:5087–5091

Johnson VA, Merrill DP, Chou T-C, Hirsch MS (1992): Human immunodeficiency virus type 1 (HIV-1) inhibitory interactions between protease inhibitor Ro 31-8959 and zidovudine, 2′,3′-dideoxycytidine or recombinant interferonαA against zidovudine-sensitive or -resistant HIV-1 *in vitro*. *J Infect Dis* 166:1143–1146

Johnson WH, Roberts NA, Borkakoti N (1987): Collagenase inhibitors: Their design and potential therapeutic use. *J Enzyme Inhib* 2:1–22

Kennedy AJ, Cline A, Ney UM, Johnson WH, Roberts NA (1987): The effect of a peptide aldehyde reversible inhibitor of elastase on a human leucocyte elastase-induced model of emphysema in the hamster. *Eur J Respir Dis* 71:472–478

Kohl NE, Emini EA, Schleif WA, Davis LJ, Heimbach JC, Dixon RAF, Scolnick EM, Sigal IS (1988): Active human immunodeficiency virus protease is required for viral infectivity. *Proc Natl Acad Sci USA* 85:4686–4690

Kramer RA, Schaber MD, Skalka AM, Ganguly K, Wong-Staal F, Reddy EP (1986): HTLV-III *gag* protein is processed in yeast cells by the virus *pol*-protease. *Science* 231:1580–1584

Krohn A, Redshaw S, Ritchie JC, Graves BJ, Hatada MH (1991): Novel binding mode of highly potent HIV-proteinase inhibitors incorporating the (R)-hydroxyethylamine isostere. *J Med Chem* 34:3340–3342

Le Grice SFJ, Mills J, Mous J (1988): Active site mutagenesis of the AIDS virus protease and its alleviation by trans complementation. *EMBO J* 7:2547–2553

Lightfoote MM, Coligan JE, Folks TM, Fauci AS, Martin MA, Venkatesan S (1986): Structural characterization of reverse transcriptase and endonuclease polypeptides of the acquired immunodeficiency syndrome retrovirus. *J Virol* 60:771–775

Martin JA, Redshaw S (1989): Amino acid derivatives. Eur Pat Appl 0,432,695

Martin JA, Bushnell DJ, Duncan IB, Dunsdon SJ, Hall MJ, Machin PJ, Merret JH, Parkes KEB, Roberts NA, Thomas GJ, Galpin SA, Kinchington D (1990): Synthesis and antiviral activity of monofluoro and difluoro analogues of pyrimidine deoxyribonucleosides against human immunodeficiency virus (HIV-1). *J Med Chem* 33:2137–2145

Martin JA, Mobberley MA, Redshaw S, Burke A, Tyms AS, Ryder TA (1991): The inhibitory activity of a peptide derivative against the growth of simian immunodeficiency virus in C8166 cells. *Biochem Biophys Res Commun* 176:180–188

Mous J, Heimer EP, Le Grice SFJ (1988): Processing protease and reverse transcriptase from human immunodeficiency virus type 1 polyprotein in *Escherichia coli*. *J Virol* 62:1433–1436

Navia MA, Fitzgerald PMD, McKeever BM, Leu C-T, Heimbach JC, Herber WK, Sigal IS, Darke PL, Springer JP (1989): Three-dimensional structure of aspartyl protease from human immunodeficiency virus HIV-1. *Nature (London)* 337:615–620

Overton HA, McMillan DJ, Gridley SJ, Brenner J, Redshaw S, Mills J (1990): Effect of two novel inhibitors of the human immunodeficiency virus protease on the maturation of the HIV *gag* and *gag-pol* polyproteins. *Virology* 179:508–511

Pearl LH, Taylor WR (1987): A structural model for the retroviral proteases. *Nature (London)* 329:351–354

Ratner L, Haseltine W, Patarca R, Livak KJ, Starcich B, Josephs SF, Doran ER, Rafalski JA, Whitehorn EA, Baumeister K, Ivanoff L, Petteway SR Jr, Pearson

ML, Lautenberger JA, Papas TS, Ghrayeb J, Chang NT, Gallo RC, Wong-Staal F (1985): Complete nucleotide sequence of the AIDS virus, HTLV-III. *Nature (London)* 313:277–284

Rich DH, Green J, Toth MV, Marshall GR, Kent SBH (1990): Hydroxyethylamine analogues of the p17/p24 substrate cleavage site are tight-binding inhibitors of HIV protease. *J Med Chem* 33:1285–1288

Rich DH, Sun C-Q, Prasad JVNV, Pathiasseril A, Toth MV, Marshall GR, Clare M, Mueller RA, Houseman K (1991): Effect of hydroxyl group configuration in hydroxyethylamine dipeptide isosteres on HIV protease inhibition. Evidence for multiple binding modes. *J Med Chem* 34:1222–1225

Roberts NA, Martin JA, Kinchington D, Broadhurst AV, Craig JC, Duncan IB, Galpin SA, Handa BK, Kay J, Krohn A, Lambert RW, Merrett JH, Mills JS, Parkes KEB, Redshaw S, Ritchie AJ, Taylor DL, Thomas GJ, Machin PJ (1990): Rational design of peptide-based HIV proteinase inhibitors. *Science* 248:358–361

Roberts NA, Craig, JC, Duncan IB (1992): HIV proteinase inhibitors. *Biochem Soc Trans* 20:513–516

Sanchez-Pescador R, Power MD, Barr PJ, Steimer KS, Stempien MM, Brown-Shimer SL, Gee WW, Renard A, Randolph A, Levy JA, Dina D, Luciw PA (1985): Nucleotide sequence and expression of an AIDS-associated retrovirus (ARV-2). *Science* 227:484–492

Toh H, Ono M, Saigo K, Miyata T (1985): Retroviral protease-like sequence in the yeast transposon. *Nature (London)* 315:691–692

Veronese F Di M, Copeland TD, De Vico AL, Rahman R, Oroszlan S, Gallo RC, Sarngadharan MG (1986). Characterization of highly immunogenic p66/p51 as the reverse transcriptase of HTLV-III/LAV. *Science* 231:1289–1291

Wong-Kai-In P, Parkes KEB, Kinchington D, Galpin S, Hope AL, Roberts NA, Martin JA, Merrett JH, Machin PJ, Thomas GJ (1991): Biological and biochemical studies on Ro 31–6840 (2′βF ddC), a dideoxynucleoside analogue active against human immunodeficiency virus Type 1 (HIV-1). *Nucleosides Nucleotides* 10:401–404

7

A New Approach to Antiviral Chemotherapy: Intervention in Viral Gene Expression by HIV Tat Antagonists

MING-CHU HSU AND STEVE TAM

Introduction

One area in which molecular virology has made major advances seems largely to have escaped the attention of pharmacologists. This is the study of how viruses regulate their gene expression. Advances in this area have been due to the fact that, being parasites of cells, viruses provide an easy handle for study of eukaroytic gene regulation in an otherwise extremely complex system. Viral genes can be amplified and modified and are easily identified in host cells. Consequently, viruses have been employed to study various steps in the regulation of gene expression, including DNA amplification, transcription, posttranscription, and translation. Such studies have provided tremendous insight into gene regulation in general. Of particular interest is the major regulation at the transcriptional level.

The discovery of adenovirus E1A protein as a viral transcriptional transactivator in 1981 revealed the complexity of viral gene transcription (Nevins, 1981). In addition to the cellular transcription machinery which the virus utilizes, adenovirus encodes a protein which temporally regulates the viral transcription. This pattern of viral transcriptional regulation is not unique to adenovirus. Viral transcriptional transactivators were soon found for pseudorabies virus (Green et al., 1983), herpes simplex virus (HSV) (Gelman and Silverstein, 1985), human immunodeficiency virus (HIV) (Arya et al., 1985; Sodroski et al., 1985), and, more recently, human hepatitis B virus (HBV) (Colgrove et al., 1989) and human cytomegalovirus (HCMV) (Chang et al., 1989; Cherrington and Macarski, 1989; Stenberg et al., 1989). These viral regulatory proteins are unique to each virus and without known cellular homologues.

The Search for Antiviral Drugs
Julian Adams and Vincent J. Merluzzi, Editors
© 1993 Birkhäuser Boston

Therefore, possibilities exist to target viral regulatory genes for the development of antivirals. Such possibilities were not explored until the emergence of HIV-1 and the devastating disease associated with the virus. In the short period of 3–4 years, scientists at Hoffmann–La Roche have made progress in this area and proven that it is a viable approach for antiviral development.

HIV Tat as an Antiviral Target

Similar to the situation seen with adenovirus and herpesvirus, HIV-1 employs a virally encoded protein, Tat (transactivator of transcription), to control its gene expression. In 1985, through deletion mapping of the HIV-1 genome, two laboratories independently discovered that the *tat* gene of HIV-1 is required for productive viral replication (Arya et al., 1985; Sodroski et al., 1985). The viral transcriptional promoter is situated within its long terminal repeat (LTR) sequence which flanks both ends of an integrated proviral DNA (Figure 1). Gene transcription of HIV-1 can be divided into two stages: basal transcription occurring before the synthesis of Tat, and the amplified transcription which is a result of Tat action. The basal transcription in CD4+ T lymphocytes follows T-cell activation, which causes, among other cellular events, activation of the cellular transcription factor NF-κB that recognizes and binds to a sequence within the LTR upstream of the transcription start site (Nabel and Baltimore, 1987). Other cellular transcription factors involved in the basal transcription are RNA polymerase II, the TATA factor (also called TFIID), the SP1 and cellular transcription factors known to be associated with Pol II and TATA factor (e.g., TFIIA, -B, -E, -F, and -G) (Figure 2). Thus, the extremely low basal transcription from the HIV-1 LTR promoter involves constitutive as well as inducible cellular factors. The basal transcription produces only the viral regulatory proteins Tat, Rev, and Nef encoded by the doubly-spliced transcript (Figure 1). This is due to Rev being required for export of the full-length primary transcript (encoding Gag and Pol) and the singly-spliced transcript (encoding Env) from the nucleus to the cytoplasm (Malim et al., 1989; Felber et al., 1989; Hammarskjold et al., 1989). After Tat is synthesized, the protein targets its cis-acting sequence, TAR, within the LTR sequence and greatly amplifies viral transcription. With Rev present, Gag, Pol, and Env mRNAs are transported to the cytoplasm and their proteins are made. Mature virions are released from the cells, and a new round of infection begins.

The importance of Tat in productive HIV-1 replication is most clearly demonstrated with mutant HIV-1 defective in the translation initiation codon of Tat. When the mutant virus was complemented with recom-

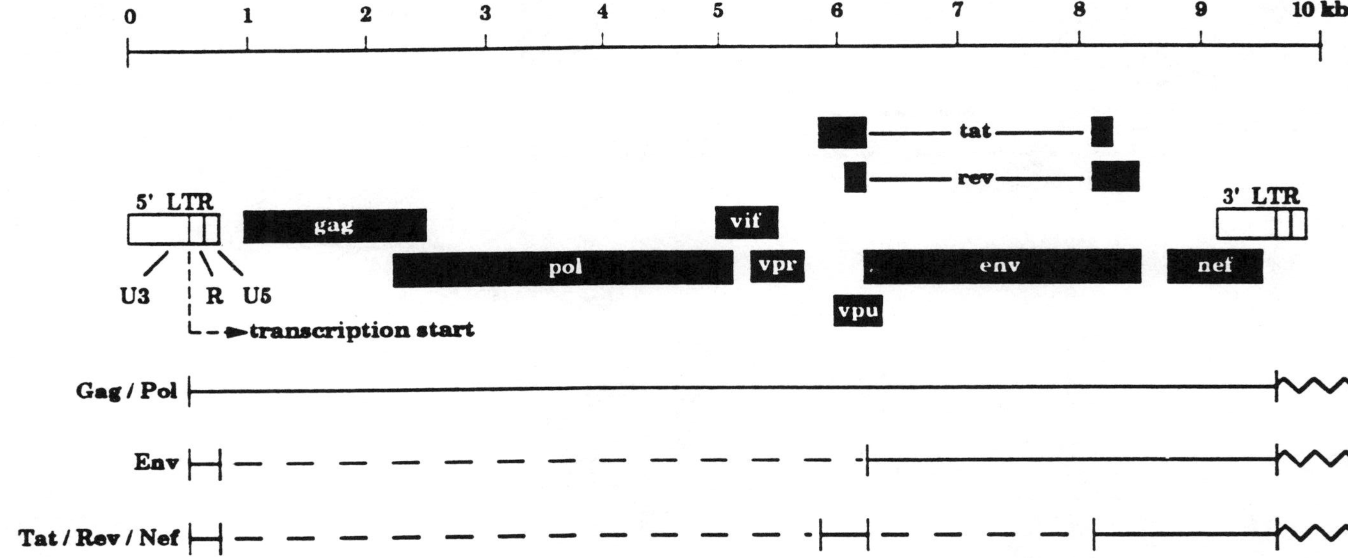

FIGURE 1. Genomic structure of HIV-1. An integrated proviral DNA is flanked by the 5′ and 3′ LTR sequence, each consisting of U3, R, and U5 segments. Transcription initiates at the 5′ end of the R segment. Primary and spliced transcripts (dashed lines designate introns) with their correspondent gene products are shown.

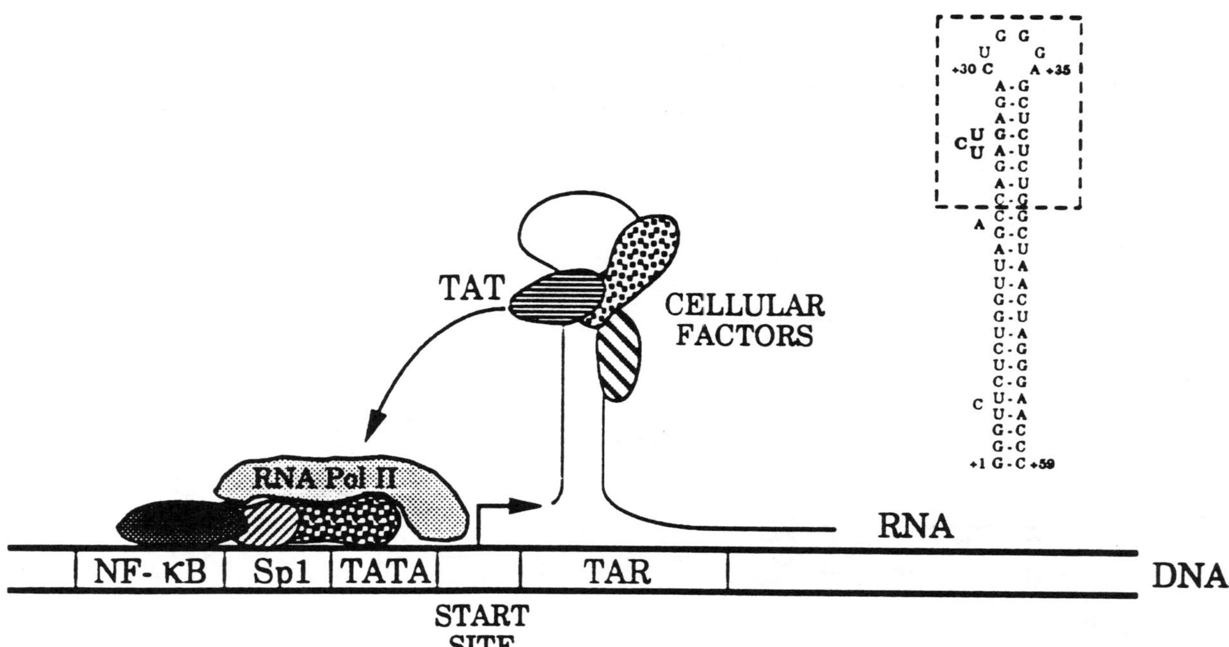

FIGURE 2. Cellular transcription factors involved in HIV-1 LTR directed transcription. TAR RNA forms a stable stem-loop structure. Tat binds to the bulge (UCU) in the stem.

binant Tat, more than a 10,000-fold increase in viral replication was found (Feinberg et al., 1991). The molecular mechanism of Tat-mediated transcriptional activation is not clearly understood. Tat is a small, basic protein (86–101 amino acids long, depending on the viral strain). Its target sequence, TAR, is downstream of the transcription start site (mapped at $+19$ to $+44$ nucleotides from the $+1$ start site) (Jakobovits et al., 1988; Hauber and Cullen 1988; Selby et al., 1989). TAR RNA forms a stem-loop structure (Feng and Holland, 1988), and in vitro studies have shown that Tat binds to a bulge in the stem (Dingwall et al., 1989, 1990; Roy et al., 1990; Weeks et al., 1990; Cordingley et al., 1990) (Figure 2). Several cellular factors have also been shown to interact with the TAR RNA (Gatignol et al., 1989, 1991; Gaynor et al., 1989; Marciniak et al., 1990; Wu et al., 1991; Sheline et al., 1991). Studies using heterologous gene constructs (HIV-1 LTR-driven indicator gene expression) showed that Tat facilitates elongation and might also increase initiation of transcription (Kao et al., 1987; Rice and Mathews, 1988; Laspia et al., 1989; Selby et al., 1989; Feinberg et al., 1991; Kessler and Mathews, 1991) (Figure 3). In the absence of Tat, the transcription complex initiated at the start site either pauses or drops off as it travels down the DNA template; thus only incomplete transcripts are found. In the

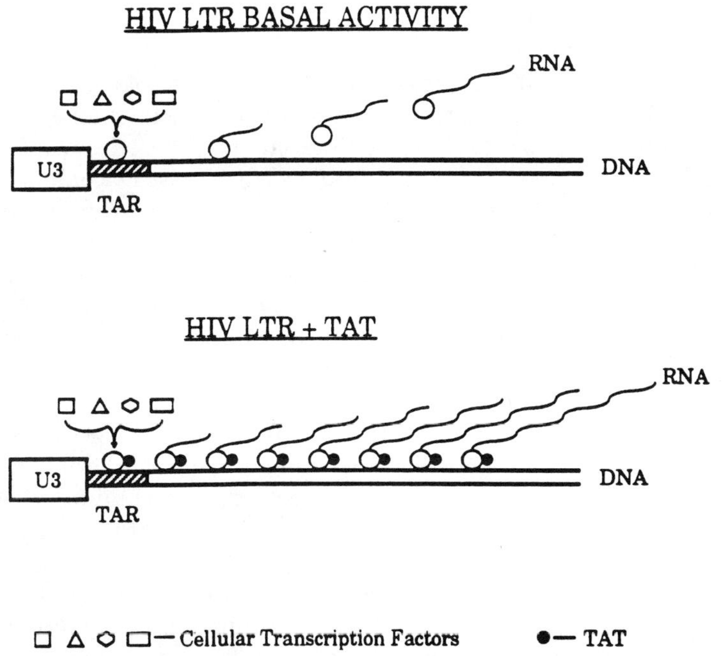

FIGURE 3. Tat stabilizes elongation and could also increase initiation of transcription from the HIV-1 LTR.

presence of Tat, the transcription complex is stabilized and can travel through the template. Thus, Tat could have a pronounced effect on a long transcript such as the HIV-1 primary transcript, which is ~10 kilobases long. Tat might also increase the loading rate of the transcription complex at the start site. How Tat, together with cellular factors, accomplishes these tasks is the subject of intensive investigation in many laboratories.

With its requirement for productive HIV-1 replication, as well as the apparent lack of a homologous cellular protein, Tat is an attractive target for the development of antivirals. However, since Tat most likely functions in concert with cellular factor(s) in transcription complexes, the challenge has been to design an assay that could distinguish a specific Tat antagonist from cytotoxic agents.

Design of a High-Flux Screen for Tat Antagonists

Before the emergence of HIV as a public health concern, Hoffmann–La Roche had a moderate amount of activity in antiviral research. At the end of 1986, the company launched a concerted international effort for the development of anti-HIV agents. With the existing interest and expertise, Roche scientists at Nutley, NJ, began to design an assay for Tat antagonists.

Since the molecular mechanism of Tat action was not understood and still proves to be elusive, it was clear that an in vitro functional assay for Tat would not be possible for some time and a cell-based transcriptional transactivation assay should be used. Such an assay could then be employed to screen the Roche compound repository or microbial broths. We reasoned that a successful assay should possess three essential characteristics: sensitivity, high-flux capacity, and reproducibility.

We serendipitously acquired a convenient indicator gene, secreted alkaline phosphatase (SeAP). The gene was derived from the membrane-bound placental alkaline phosphatase, which was cloned and investigated by Drs. J. Berger and S. Udenfriend of Roche Institute of Molecular Biology and Dr. B. Cullen of Hoffmann–La Roche. Deletion of the sequence encoding the C-terminal membrane-bound domain of the protein caused secretion of the enzyme into culture medium (Berger et al., 1988). Activity of the enzyme could then be measured with a simple and fast colorimetric assay. With the SeAP gene put under the control of the HIV-1 LTR promoter, SeAP activity in the culture media is a measure of Tat transactivation of the promoter (Figure 4). To determine the cytotoxicity of a test compound, the SeAP gene was put under the control of the Rous sarcoma virus (RSV) LTR, which is not responsive to Tat. A cytotoxic agent or an agent that inhibits alkaline phosphatase

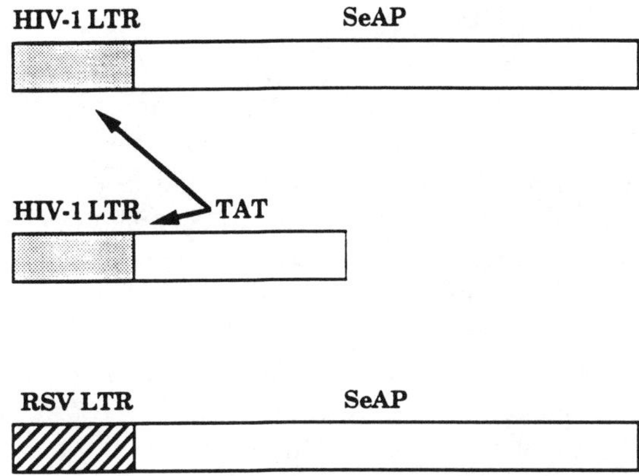

FIGURE 4. Gene constructs employed in screening assay for Tat antagonists. The plasmid vector pBC12 was used for expression of the gene constructs (Berger et al., 1988).

activity would inhibit SeAP activity produced from both gene constructs. Thus, in addition to the high-flux capacity provided by the indicator gene SeAP, the assay would discriminate specific inhibitors of HIV-1 LTR promoter activity from general cytotoxic agents.

However, with the lack of a known Tat antagonist, whether a given assay design could identify an antagonist could not be ascertained. To increase our chance of success, the assay was designed to achieve a high sensitivity. To that end, *tat* gene expression was put under the control of the HIV-1 LTR (Figure 4) and the assay was performed under conditions in which SeAP production was in linear response to Tat produced in the cells (Hsu et al., 1991). An antagonist to Tat would therefore reduce SeAP as well as *tat* gene expression, a situation analogous to viral gene expression from the HIV genome. Such a design greatly enhanced the sensitivity of the assay and also eliminated the problem of false positives resulting from inhibition of *tat* expression driven by other promoters. To control the quantity of Tat in cells, we chose to perform a transfection assay rather than a less tedious assay employing a constitutive cell line. However, the inherent variability in efficiency of transfecting cells in monolayer needed to be overcome. We then developed a protocol in which large numbers of trypsinized COS cells were transfected in suspension and then plated into 96-well microtiter plates. COS cells amplified the input plasmids containing the SV40 replication origin and high SeAP activity was detected in the culture media. We were able to achieve ±10% reproducibility with such a transfection protocol. With a semiautomated procedure, a person was able to test 200–300 compounds per week with multiple concentrations and parallel cytotoxicity control assays. This was equivalent to 4000–6000 alkaline phosphatase assays per week.

The Lead Compound

One month into our screening of the Roche compound repository, in November 1987, we were pleasantly surprised to come across Ro 5–3335 (Figure 5). The compound specifically inhibited SeAP gene expression controlled by the HIV-1 LTR promoter in a dose-dependent manner without inhibiting expression of the SeAP gene driven by the RSV LTR (Figure 6A) (Hsu et al., 1991). The potency of the compound in these transfection assays was moderate, with a 50% inhibition (IC_{50}) at 1-μM concentration and a 80–90% inhibition, without cytotoxicity, at 50-μM concentration. However, when the compound was tested with cell lines that carried gene constructs HIV-1 LTR/SeAP and HIV-1 LTR/Tat in their chromosomes and expressed SeAP constitutively (a situation analogous to integrated HIV-1 provirus), the compound showed a greater potency, with an IC_{90} < 1 μM (Figure 6B). Concomitant reduction in enzyme activity and SeAP-specific mRNA (measured by Northern blotting) was observed, an indication that Ro 5–3335 inhibited SeAP production at the level of gene transcription.

Dr. David Volsky (St. Luke's/Roosevelt Hospital, New York City) collaborated with us to design anti-HIV testing for the compound. Collaboration was later expanded to include a partnership in a National Cooperative Drug Discovery Group. Approximately 1 year later, with an increasing number of analogue compounds to be evaluated, Dr. Douglas Richman (Veterans Administration Medical Center, University of California, San Diego) joined our effort and greatly increased our testing capacity for different cell types, viral strains, and testing protocols. More recently, his investigation of viral resistance to this class of compounds has added invaluable insight into the mechanism of action.

A low antiviral potency was found when Ro 5–3335 was first tested in

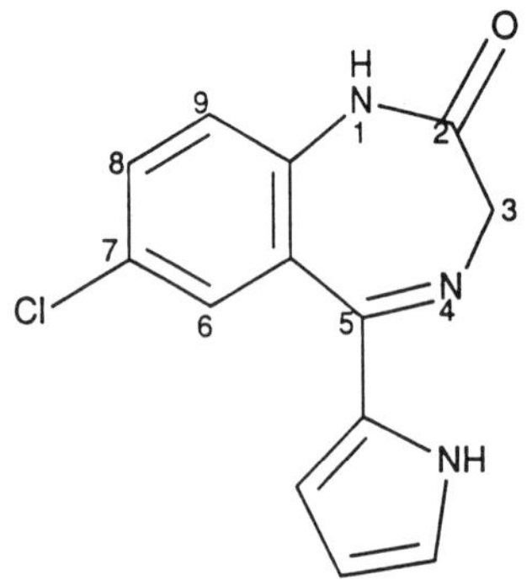

FIGURE 5. The lead compound Ro 5–3335.

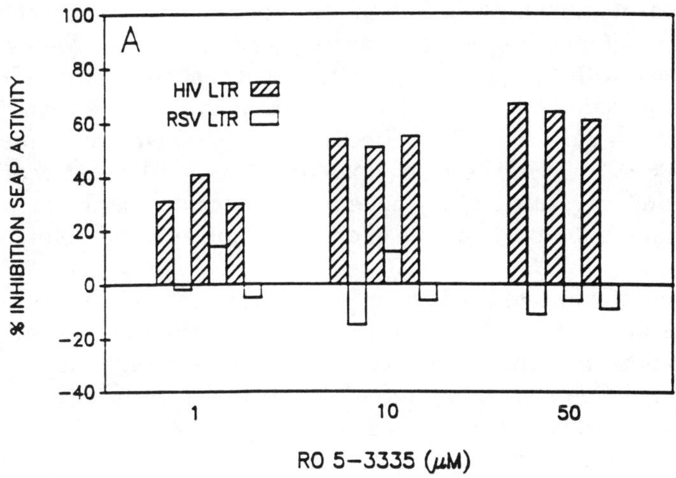

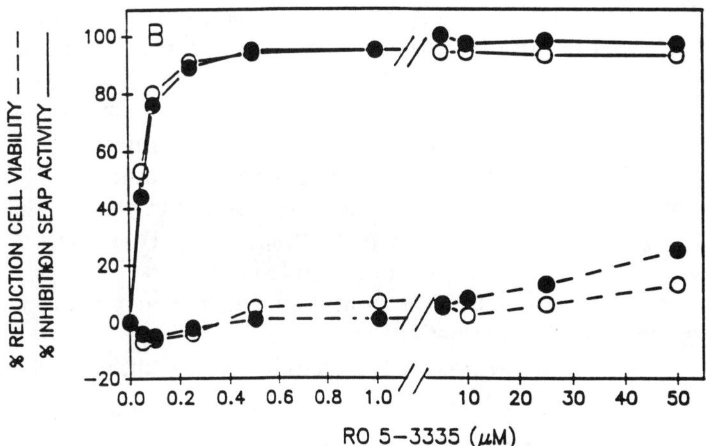

FIGURE 6. Ro 5–3335–inhibited SeAP gene expression controlled by HIV-1 LTR (Hsu et al., 1991): (A) three repeats of transfection assay; (B) assays with two constitutive cell lines. Cell viability was measured by the tetrazolium-base MTT [3-(4,5-dimethylthiazol-2-yl)-2,5-diphenyl-2*H*-tetrazolium bromide] assay (Denizot and Lang, 1986).

assays that employed an extremely high multiplicity of infection (MOI) and cell death as the end-point measurement in cells supporting high HIV-1 replication (e.g., HTLV-1-transformed HUT78, MT4, and MT2 cells). We quickly realized that such an assay protocol, although proven useful for RT inhibitors, might not be optimal for testing a Tat antagonist. As a regulatory protein, a minute quantity of Tat is required in infected cells to support viral replication. When a high MOI is used, a very potent inhibitor will be required to stop viral replication and a potential Tat inhibitor could be missed. A titration of the MOI was then performed with CEM and Jurkat cells in acute infection (drug addition at time of infection). At low MOIs that result in no more than 50% cell death in 3–4 days, Ro 5–3335 effectively inhibited viral replication, with an IC_{50} = 0.1–1 μM. It became apparent that anti-HIV potency of a given compound, regardless of its mode of action, could vary owing to MOI, viral strain, cell type, and time of drug addition and end-point measurement.

Chronically infected cells then provided a clear demonstration of Ro 5–3335's mode of action. Unlike acute infection, there are no new rounds of infection in chronically infected cells. This presumedly is due to downregulation of the viral receptor CD4 molecules at the cell surface (see Table 1; Hoxie et al., 1985, 1986; McDougal et al., 1985). Ro 5–3335 effectively inhibited viral replication in such a chronic culture by reducing viral transcription from integrated provirus, as demonstrated by the reduction of cell-associated viral RNA and antigen (Figure 7) (Hsu et al., 1991). As expected, a reverse transcriptase inhibitor, azidothymidine (AZT), could not inhibit HIV-1 replication in this chronic culture. To further demonstrate the mode of action of Ro 5–3335, Jurkat cells were transfected with an infectious HIV-1 DNA clone. Within 24 hours after transfection, Ro 5–3335 was more effective in suppressing viral p24 antigen production than were soluble CD4 or AZT, and a combination treatment with Ro 5–3335 and soluble CD4 resulted in an antiviral effect similar to that of Ro 5–3335 alone (Hsu et al., 1991). These data showed that Ro 5–3335 inhibited viral replication from proviral DNA and ruled out the possibility that the compound intervened in viral replication prior to proviral DNA formation.

Ro 5–3335 has a broad spectrum of activity against different strains of

TABLE 1. Expression of HIV-1 proteins and surface CD4 in chronically infected CR10/N1T cells cultured in the presence of Ro 5–3335[a,b]

Ro 5–3335 dose (μM)	Cell viability (% living cells)	Expression of HIV-1 proteins		Surface CD4 expression (% positive cells)
		IF-positive cells (%)	p24 (ng/10^6 cells)	
0	95	66	2,067	0.7
0.1	95	50	1,871	1.3
1	93	15	400	8.5
5	93	6	45	24.8
10	90	2	33	29.1

[a]Reprinted with permission of the *Journal of Virology* from Shahabuddin M. et al. (1992): Restoration of cell surface CD4 expression in human immunodeficiency virus type 1–infected cells by treatment with a Tat antagonist. *J Virol* 66:6802–6805.
[b]Assays were performed after 6-day drug treatment. *Key:* IF = immunofluorescent antibodies.

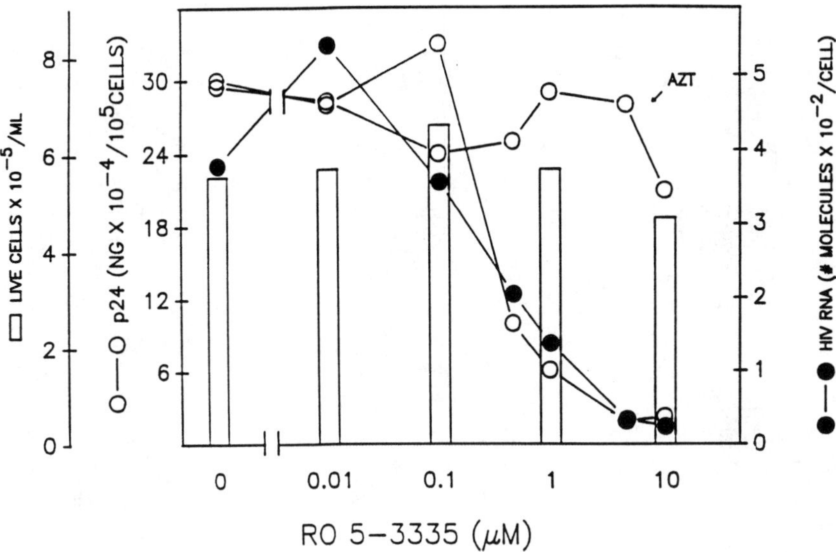

FIGURE 7. Ro5–3335 inhibited HIV-1 replication in CR10 cells chronically infected with strain N1T virus. Reprinted with permission of the AAAS from Hsu MC et al. (1991): Inhibition of HIV replication in acute and chronic infections *in vitro* by a Tat antagonist. *Science* 254:1799–1802. © AAAS.

HIV-1 (3B, Bru, BAL/85, N1T, and RF) in several cell lines (CEM, Jurkat, H9, JM, C8166, and HT4-6C) and in primary peripheral blood lymphocytes and macrophages. Similar potencies were found (IC_{50} = 0.1–1 μM; IC_{90} = 1–3 μM) with different end-point measurements [p24 ELISA (enzyme-linked immunosorbent assay), plaque reduction, syncytium inhibition, antibody staining of infected cells, and viral RNA quantitation]. The compound was also active against HIV-2 (strains ROD and UC-1) and AZT-resistant clinical isolates.

Chemistry

Because of our experience in benzodiazepine chemistry at Roche, the discovery of Ro 5–3335 as a Tat antagonist came as a very pleasant surprise. However, after a computer search of our in-house compound repository and subsequent testing of analogue compounds, we realized that the Tat-inhibitory activity is not a general property of either a benzodiazepine or a pyrrole. Simple methylation of either the pyrrole or the amide NH or both led to a complete loss of activity, indicating to us the biological activity resides in the entire molecule of Ro 5–3335. The decision to launch a full-scale chemical synthesis program was made immediately after the compound was shown to reduce cell-associated viral RNA in chronically infected cells. The data was obtained with a technique called liquid RNA–RNA hybridization/target cycling, which was under development by GeneTrak System (Framingham, Massachusetts) at that time. This assay has the sensitivity of quantitating (without amplification) 1–10 pg of HIV RNA in 50–100 chronically infected cells (Volsky et al., 1990). Using this assay, Dr. Volsky, our collaborator, showed that Ro 5–3335 caused a reduction of HIV RNA with an IC_{50} between 0.1 and 1 μM (Figure 7).

For a systematic study of the structure–activity relationship, six regions of the molecule of Ro 5–3335 were targeted for structural modifications. These included changes at the N-1–C-2 amide group, substitution at the C-3 carbon, modifications of the N-4–C-5 imine group, replacements of the pyrrole moiety at C-5, substitutions on the benzene ring, and complete replacements of the benzodiazepine skeleton. Over a period of 15 months, about 300 analogues were made and tested in the anti-Tat assay, a testimony to the enormous enthusiasm of the chemists and biologists involved.

Isosteric replacement of the amide oxygen with sulfur led to a reduction in anti-Tat activity, whereas in the amidine series the activity was clearly related to the size of the substituent(s). Bulky or aromatic groups were not allowed at this position. The best compounds in this series were Ro 24–7429, Ro 24–8516, Ro 24–8609, Ro 24–8852, and Ro 24–9456 [Figure 8(a)]. Interestingly, acetylation of the hydrazine group to give the hydrazide (Ro 24–8739) completely abolished the activity. Other more drastic changes, such as Ro 24–7787, Ro 24–8511, Ro 22–1269, Ro 24–8762, and Ro 24–8776 all resulted in a complete loss of anti-Tat activity.

Incorporations of a heteroatom (OH, OAc, OMe or NH_2) at the C-3 position were detrimental to the activity. However, of the two optically active 3-methyl derivatives, only the *R*-isomer possesses anti-Tat activity, whereas its enantiomer is totally inactive. This structure–activity relationship was exactly opposite to what was observed in the central nervous system (CNS) activity of the diazepam derivatives. Additional derivatizations at the C-3 position also revealed a poor tolerance for bulky or aromatic substituents.

Replacement of the pyrrole moiety at the C-5 position with larger heterocyclics (e.g., pyridine, pyrimidine, indole, benzimidazole) or carbacyclic rings (e.g., cyclopentyl or phenyl) abolished anti-Tat activity. Small non-nitrogen-containing heterocycles, such as thiophene, furan, thiazole, and isothiazole, at this position also resulted in inactive compounds. The most interesting findings resulted from substitutions with other nitrogen-containing 5-membered heterocycles [Figure 8(b)]. The 4-substituted pyrazole Ro 25–2483, 3-substituted pyrrole Ro 25–0984, and 1,2,3-triazole Ro 25–0982 all exhibited high anti-Tat activity. The 4-substituted imidazole Ro 25–1429 and 3-substituted pyrazole Ro 24–9257 showed only moderate anti-Tat activity, whereas the 2-substituted imidazole Ro 24–0334 and 3-substituted 1,2,4-triazole Ro 25–0500 were inactive. Based on the structures revealed by X-ray crystallography, we were not able to explain these differences in the activity. On the other hand, the electrostatic potential maps of these compounds were clearly different and this may be the contributing factor for the wide range of activity observed.

For the benzene ring, it was clear only the substitution at C-7 is allowed. Substitutions at positions 6, 8, or 9 all led to a loss of activity. The optimal substituents at C-7 were the methyl, the fluoro, and the chloro groups.

During the course of this program, many analogues were selected for testing of anti-HIV activity in cell cultures. The anti-Tat activity of each compound correlated well with the result of antiviral assay when different assay protocols were performed in Dr. D. Volsky's and Dr. D. Richman's laboratories. All of the analogues with anti-Tat activity also showed anti-HIV activity, and none of inactive compounds in the anti-Tat assay exhibited any anti-HIV activity.

Identification of a Clinical Candidate

Ro 5–3335 was synthesized at Roche in 1959 for the CNS program but was found to be inactive. Its affinity to the benzodiazepine receptor in the rat CNS was less than 1% of that of diazepam. This compound was 85% bioavailable and had a half-life of 2 hours in dogs. With a 1 mg/kg dose, a 0.8-μM plasma concentration was achieved. Unfortunately, when the compound was tested in rats, a mild-to-severe nephrotoxicity caused by yellow casts deposited in renal tubules was discovered and there was tissue discoloration caused by a yellow metabolite, the α-hydroxypyrrole.

A decision was then made to test potential development candidates in short-term (5-day) toxicity studies in rats. At a dosage of 100 mg/kg per day, several analogues did not show nephrotoxicity and had reduced tissue discoloration (Ro 24–7429, Ro 24–8609, and Ro 24–8516) or none (Ro 25–2483 and 24–9257). Based on a more favorable in vitro safety window and animal toxicity data, Ro 24–7429 was chosen to move on to longer term toxicity studies. The compound has an anti-Tat and an antiviral activity comparable to that of Ro 5–3335. It has a good bioavailability in dogs (~35%) and a low affinity for the CNS benzodi-azepine receptors (~0.1% of diazepam). Our efforts in toxicology moved to a rapid conclusion when no further toxic liabilities were found in rats after 4-week exposure of up to 100 mg/kg per day and, more

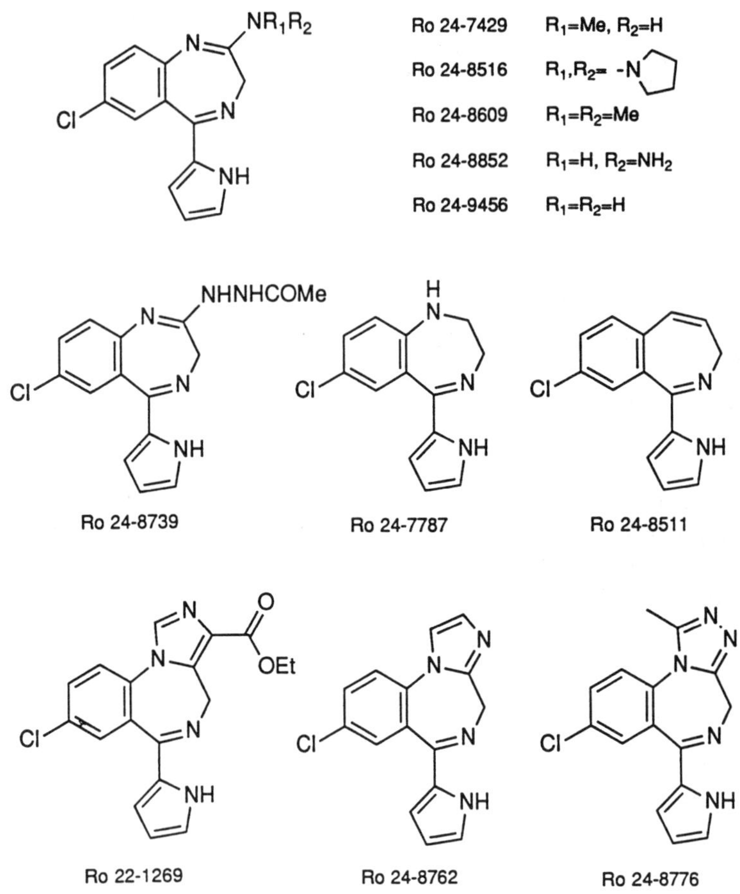

Ro 24-7429 R_1=Me, R_2=H

Ro 24-8516 R_1,R_2= -N⟨⟩

Ro 24-8609 R_1=R_2=Me

Ro 24-8852 R_1=H, R_2=NH$_2$

Ro 24-9456 R_1=R_2=H

Ro 24-8739

Ro 24-7787

Ro 24-8511

Ro 22-1269

Ro 24-8762

Ro 24-8776

FIGURE 8. (a) Analogues of Ro 5-3335 in the N-1–C-2 series.

encouragingly, the compound was well tolerated and found not to cause tissue discoloration in squirrel monkeys with 4-week dosing of 600 mg/kg per day. Following an intensive preclinical development, the compound was first tested in a dose-ranging study in humans in March 1991.

Molecular Mechanism of Action of Ro 5–3335

The complexity of Tat-mediated transcriptional activation, which involves the cellular transcription machinery, implies that multiple targets exist for antiviral development as well as potential side effects of an antagonist. Specific Tat antagonists could possibly include compounds that break down interactions of Tat/TAR RNA, Tat/cellular factors, and TAR RNA/cellular factors or prevent nuclear localization of Tat (Figure 2).

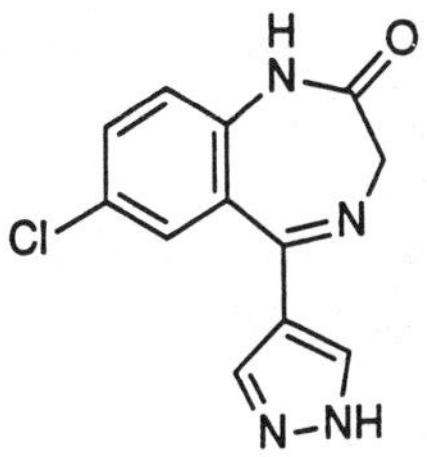

Ro 25-2483

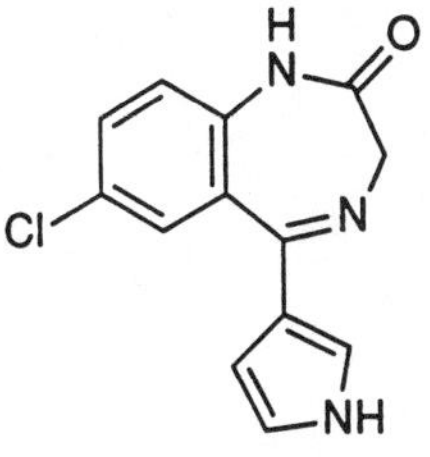

Ro 25-0984

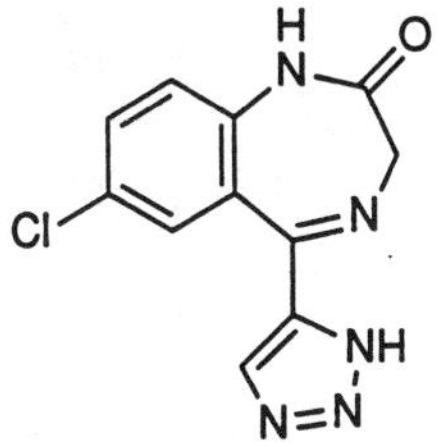

Ro 25-0982

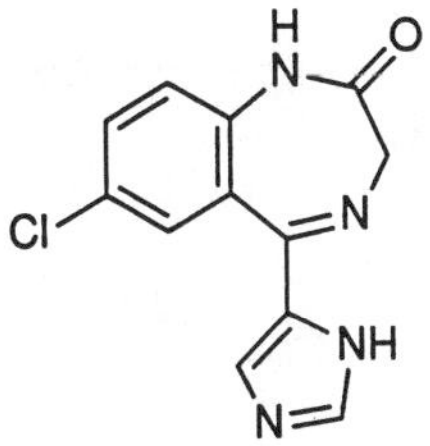

Ro 25-1429

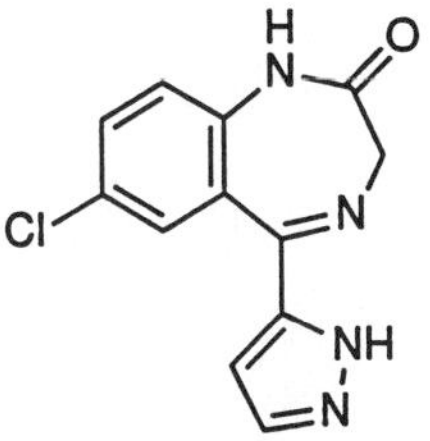

Ro 24-9257

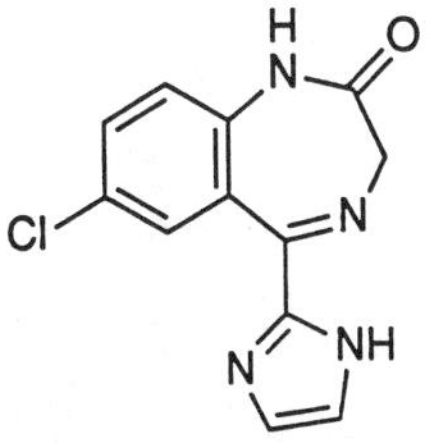

Ro 24-0334

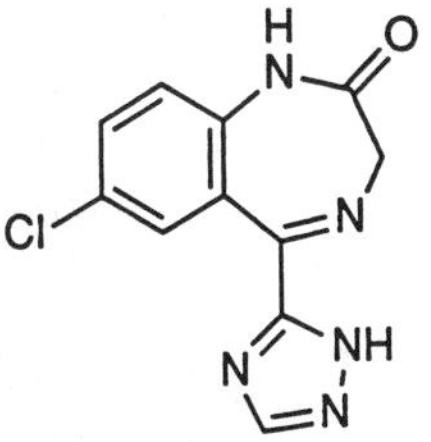

Ro 25-0500

FIGURE 8 (*cont.*). (b) Analogues of Ro 5–3335 in the C-5 series.

Using radiolabeled and functionally active recombinant Tat, we showed that Ro 5–3335 did not prevent Tat from getting into nuclei (L. W. Slice and M. C. Hsu, unpublished observations). Since the compound did not inhibit transcription of the cellular actin gene, or a gene controlled by another viral promoter at active concentrations (Figure 6), cellular core transcription factors might not be involved in its action. We investigated this question further by examining the basal activity of the HIV-1 LTR. The studies were performed with COS cells which support a measurable basal activity of the HIV-1 LTR–driven chloramphenicol acetyltransferase (CAT) gene, a highly sensitive indicator gene. The compound did not inhibit CAT production in the absence of Tat and deletion of the TAR sequence rendered the promoter nonresponsive to the compound (Figure 9) (Hsu et al., 1991). The specificity of Ro 5–3335 for Tat and TAR was thus established.

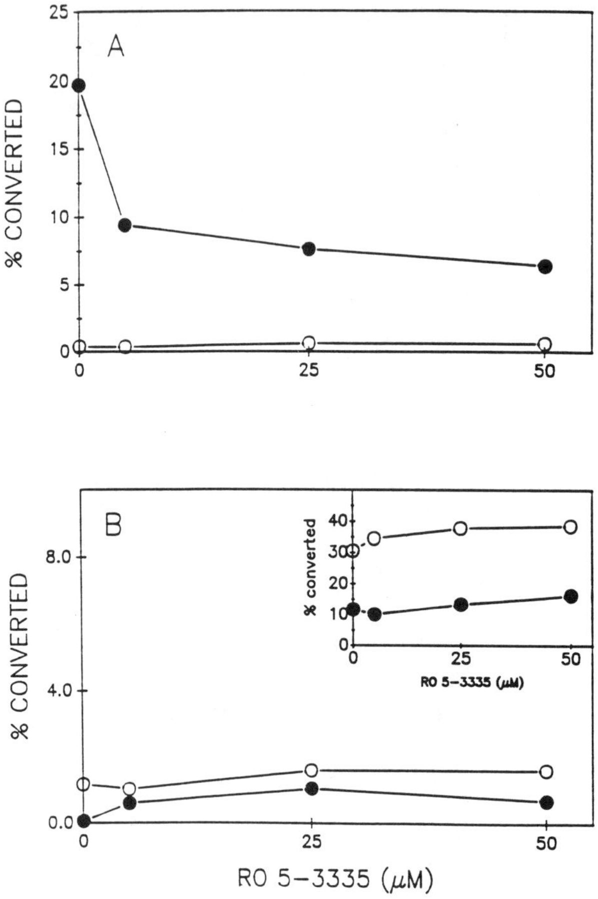

FIGURE 9. Ro 5–3335 inhibition of gene expression controlled by HIV-1 LTR is dependent on Tat and TAR. CAT gene expression was driven by the wild-type HIV LTR in (A), and by the deletion mutant (for nucleotides +9 to +77) in (B). CAT activity was expressed in percentage conversion of chloramphenicol into acetylated products by the enzyme: (•) in the presence and (○) in the absence of Tat expression. Reprinted with permission of the AAAS from Hsu MC et al. (1991): Inhibition of HIV replication in acute and chronic infections *in vitro* by a Tat antagonist. *Science* 254:1799–1802. © AAAS.

However, the foregoing observations did not rule out the possibility that Tat exerts its function through interaction with cellular factor(s) and Ro 5–3335 can break down the interaction. To address this possibility, the upstream DNA sequence recognized by the TATA, the SP1, or the NF-κB transcription factors were mutated (constructs kindly provided by Dr. D. Baltimore) and each mutant promoter was tested for its responsiveness to Ro 5–3335. The compound similarly inhibited gene expression from these mutant promoters when compared with the wild-type promoter (Hsu et al., 1991). We concluded that the action of the drug was not mediated through these cellular transcription factors upstream from the transcription start site.

Tat has been shown to bind to TAR RNA in vitro with high affinity ($K_d \cong 10^{-9}$ M) and some specificity (Dingwall et al., 1989, 1990; Roy et al., 1990; Weeks et al., 1990; Cordingley et al., 1990). Ro 5–3335 did not bind to recombinant Tat or break down the Tat–TAR RNA interaction in binding assays in vitro (D. Antelman and M. C. Hsu, unpublished observations). Although mutational analyses have demonstrated the requirement of TAT–TAR RNA interaction for Tat action, whether the in vitro binding does or does not truly reflect the situation in vivo, where Tat is situated in a multiprotein complex, remains to be investigated. Several cellular proteins have also been identified which bind to various regions of TAR RNA (Gatingnol et al., 1989, 1991; Gaynor et al., 1989; Marciniak et al., 1990; Wu et al., 1991; Sheline et al., 1991). To address the possibility that these proteins are involved in the inhibitory action of Ro 5–3335, we tested the compound with a mutant promoter (kindly provided by Dr. B. M. Peterlin) in which the TAR sequence was replaced by the MS2 operator sequence that forms a similar stem-loop RNA structure and is recognized by the phage coat protein (CP). A fusion protein of Tat and CP recognized the operator RNA and transactivated the mutant promoter with 10–20% efficiency of the wild-type promoter (Selby and Peterlin, 1990). However, Ro 5–3335 was found not to inhibit transactivation from the mutant promoter (Table 2). Thus, it appears that cellular factors which recognize TAR RNA might be involved in the action of Ro 5–3335. Our data from recombinant Tat-mediated transactivation in which Tat protein was taken up by cells from the culture medium could support such a mechanism. By varying the quantity of recombinant Tat in culture media, a dose-dependent activation of the indicator gene CAT driven by the HIV-1 LTR (in a constitutive cell line HL3T1 kindly provided by Dr. G. Pavlakis; see Felber and Pavlakis, 1988) was observed. However, at 1-μM concentration, Ro 5–3335 totally inhibited CAT expression with 2, 10, or 50 ng of Tat in the media (Figure 10). These results indicate that Tat might not be the limiting factor in order for the drug to exert its action. Direct proof that a cellular protein

TABLE 2. Effect of Ro 5–3335 on promoter activity

Promoter	Inhibition
HIV-1 (TAR)/Tat	+
HIV-1 (MS2 operator)/Tat-CP	−
RSV	−
HCMV	−

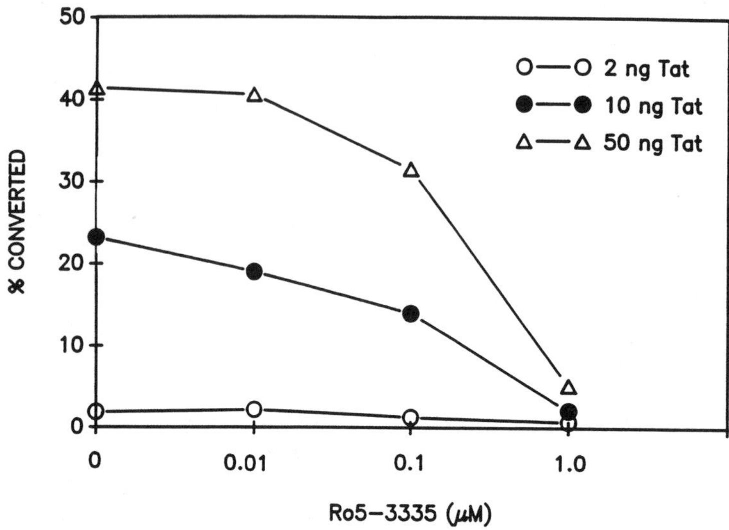

FIGURE 10. Ro 5–3335 inhibited expression of CAT gene driven by the HIV LTR when recombinant Tat protein was provided in culture media and uptaken into the cells.

is involved in the action of Ro 5-3335 will come from identification of the cellular protein and functional reconstitution in a cell-free system.

Another strong piece of evidence in support of involvement of a cellular protein in Tat and/or Ro 5-3335 action came from the lack of viral resistance to the drug. In the clinic we are confronted with the problem of viral resistance to AZT (Larder et al., 1989) and certain nonnucleoside RT inhibitors (Richman et al., 1991; Nunberg et al., 1991). The high mutation rate of the virus will continue to be a challenge both in the clinic and in drug design. If Tat indeed interacts with cellular proteins, which are invariant, and Ro 5-3335 antagonizes this interaction, we would expect that resistance to Ro 5-3335 will be slow to develop. This was found to be true for Ro 24-7429, which has undergone long-term testing in the laboratory of Dr. Douglas D. Richman. With a weekly virus passage in fresh CEM cells and a drug concentration of 10 μM which results in >95% inhibition of virus growth, viral resistance to the drug was not found after 2 years (Hsu et al., 1993). Whether this observation will hold up in the clinic or not remains to be seen.

Although resistance would develop slowly or not at all to antivirals targeted at cellular proteins or viral–cellular protein complexes, such drugs could be toxic to cells. The safety window for Ro 5-3335 was found to vary from 10 to 500, depending on cell types. The drug appeared to be more toxic to certain transformed cell lines (e.g., HUT78, MT4, and MT2). In addition to possible differences in drug penetration, it is conceivable that Tat–cellular protein interactions could vary among cell types. Thus, selection of a clinical candidate should be based on animal toxicity. We were encouraged by the finding that Ro 24–7429 had no apparent immunotoxicity in mice, rats, monkeys, or dogs at plasma concentrations in excess of the in vitro IC_{90}.

Owing to the complexity of the transcription complex, although significant progress has been made in understanding the mechanism of action of Tat, we do not yet have a model that could unify all the experimental findings. An analogy might be made with the immediate early gene product EIA of adenovirus. The gene function was discovered in 1981, yet its molecular mechanism of action only began to emerge recently. Ro 5–3335 could offer another tool for investigation of Tat action. With a concerted effort of scientists in antiviral and gene regulation research, we can expect that good progress will be made in the next few years.

Complementation by a Tat Antagonist of Other Anti-HIV Agents

Ten years into the AIDS epidemic, the pathogenesis of HIV-1 is not yet fully understood. The virus causes cell death and down-regulation of cell surface CD4 molecules in vitro (Hoxie et al., 1985, 1986; McDougal et al., 1985). Both mechanisms have the consequence of dampening host immune responses. The number of infected CD4+ T cells in patients, however, cannot fully account for the depletion of T cells. On the other hand, several retrospective studies have clearly demonstrated the association of disease progression with viral load in patients (Coombs et al., 1989; Ho et al., 1989; Schnittman et al., 1990). Reduction of viral load by preventing viral replication will continue to be the main and most valid approach for AIDS therapy. Since the size of the viral reservoir in patients is presently not clear to us and it is not known how long infected cells can persist before dying, an important question regarding the therapeutic approach is whether preventing new rounds of infection (e.g., by a reverse transcriptase inhibitor, a protease inhibitor, or soluble CD4) would be sufficient to reduce viral load in patients. Clinical experience with AZT has indicated that throughout the course of AZT treatment and before the appearance of viral resistance, there is little reduction of viral load in patients' peripheral blood mononuclear cells (Ho et al., 1989). Whether this is due to the lack of in vivo potency of the drug or to the persistance of chronically infected cells is not clear at present. A drug that can prevent HIV-1 replication in chronically infected cells, such as a Tat antagonist, should be ideal for combination therapy with drugs that prevent new rounds of infection.

Another property of a Tat antagonist that is not shared by other anti-HIV agents is its capacity to restore cell-surface CD4 molecules. Down-regulation of CD4 molecules on the surface of HIV-1 infected cells is the result of intracellular viral gp 160 interaction with CD4 molecules, preventing the appearance of CD4 molecules on cell surfaces (Jabbar and Nayak, 1990; Crise et al., 1990; Bour et al., 1991). By inhibiting viral transcription and hence viral protein synthesis, Ro 5–3335, as expected, restored cell surface CD4 molecules in chronically infected cells [Table 1 (Shahabuddin et al., 1992)].

In addition to cell killing and down-regulation of cell surface CD4 molecules, in vitro studies have offered several other possible pathogenic mechanisms for HIV-1. The viral surface glycoprotein gp 120 has been shown to mediate cytotoxic T-cell killing of infected cells (Siliciano et al., 1988). The protein could also dissociate from virions or the surface of infected cells, or be released from dead cells. The free gp 120 can bind to CD4 molecules of uninfected T cells and elicit cell lysis through antibody-dependent cellular cytotoxicity (Tanneau et al., 1990). In addition, HIV-1 Tat was shown to reduce antigen-dependent T-cell proliferation (Viscidi et al., 1989) and to stimulate growth of Kaposi's sarcoma–derived cells in vitro (Ensoli et al., 1990). If these in vitro findings do exist in vivo, therapy with a Tat antagonist, which prevents viral transcription and thereby viral protein production, might have other benefits in addition to protecting chronically infected cells and preventing new rounds of infection.

Future Prospects

The ultimate test of an antiviral is in the clinic. In that regard, we have had only a few successful stories before the AIDS epidemic. On the other hand, we have never been confronted with as devastating a virus as HIV-1 and, as documented in this book, the epidemic has stimulated unprecedented efforts in antiviral research and development. The opportune development of the Tat antagonists, with their potentials distinct from other classes of anti-HIV agents, may herald a new avenue for antiviral development.

As mentioned in our introduction, many viruses pathogenic to humans employ viral transcriptional transactivators to regulate viral gene expression. These viral regulatory gene products are potential targets for antiviral development, and Roche's Tat antagonists have demonstrated the viability of such an approach. Now that development of anti-HIV agents has greatly enriched our experience and increased the repertoire of antiviral targets, scientists might well first weigh the potential benefits of each approach before embarking upon a drug discovery effort. Whether a viral transcriptional regulator is a suitable antiviral target will depend on the understanding of the disease mechanism of a particular viral infection.

Should a Tat antagonist succeed in the clinic, it would be the first drug designed to intervene in gene regulation; however, it would not be the first known to possess this mode of action. Cyclosporin A and FK506 have both been used in the clinic for some time and were recently shown to suppress immune responses by inhibiting IL-2 gene expression at the level of transcription. Both drugs block nuclear localization of a T-cell transcription factor NF-AT which is essential for IL-2 promoter activity (Flanagan et al., 1991). Anti-inflammatory activity of glucocorticoids has been shown to correlate with their inhibitory action on the transcription factor AP-1 that promotes expression of collagenase (Jonat et al., 1990). These drugs target basic cellular functions and have toxic side effects, yet have been successfully managed in the clinic. These drugs, and hopefully an HIV Tat antagonist in the near future, further support the viability of therapeutic intervention in gene regulation.

Acknowledgments

We would like to acknowledge Dr. David J. Volsky, Dr. Douglas Richman, and members of Virology, Anti-infective Chemistry, Investigative Pathology, Toxicology, and Drug Metabolism of Hoffmann–La Roche. Their contribution and inspiration made the entire discovery and development process possible. The basic research on mechanism of action of Tat and its antagonists was partly supported by postdoctoral fellowships from the National Cooperative Drug Discovery Group (AI27397).

References

Arya SK, Guo C, Josephs SF, Wong-Staal F (1985): Transactivator gene of human T-lymphotropic virus type III (HTLV-III). *Science* 229:69–73

Berger J, Hauber J, Hauber R, Geiger R, Cullen BR (1988): Secreted placental alkaline phosphatase: A powerful new quantitative indicator of gene expression in eukaryotic cells. *Gene* 66:1–10

Bour S, Boulerice F, Wainberg MA (1991): Inhibition of gp160 and CD4 maturation in U937 cells after both defective and productive infections by human immunodeficiency virus type 1. *J Virol* 65:6387–6396

Chang CP, Malone CL, Stinski MF (1989): A human cytomegalovirus early gene has three inducible promoters that are regulated differentially at various times after infection. *J Virol* 63:281–290

Cherrington JM, Macarski ES (1989): Human cytomegalovirus IE1 transactivates the alpha promoter-enhancer via an 18-base-pair repeat element. *J Virol* 63:1435–1440

Colgrove RG, Simon G, Ganem D (1989): Transcriptional activation of homologous and heterologous genes by the hepatitis B virus x gene product in cells permissive for viral replication. *J Virol* 63:4019–4026

Coombs RW, Collier AC, Allain JP, et al. (1989): Plasma viremia in human immunodeficiency virus infection. *N Engl J Med* 321:1626–1631

Cordingley MG, LaFemina RL, Callahan PL, Condra JH, Sardana VV, Graham DJ, Nguyen TM, LeCrow K, Gotlib L, Schlabach AJ, Colanno RJ (1990): Sequence specific interactions of Tat protein and Tat peptides with the transactivation-responsive sequence element of human immunodeficiency virus type 1 *in vitro. Proc Natl Acad Sci USA* 87:8985–8989

Crise B, Buonocore L, Rose JK (1990): CD4 is retained in the endoplasmic reticulum by the human immunodeficiency virus type 1 glycoprotein precursor. *J Virol* 64:5585–5593

Denizot F, Lang R (1986): Rapid colorimetric assay for cell growth and survival: Modifications to the tetrazolium dye procedure giving improved sensitivity and reliability. *J Immunol Methods* 89:271–277

Dingwall C, Ernberg I, Gait MJ, Green SM, Heaphy S, Karn J, Lowe AD, Singh M, Skinner MA, Valerio R (1989): Human immunodeficiency virus 1 Tat protein binds trans-activation-responsive region (TAR) RNA *in vitro. Proc Natl Acad Sci USA* 86:6925–6929

Dingwall C, Ernberg I, Gait MJ, Green SM, Heaphy S, Karn J, Lowe AD, Singh M, Skinner MA (1990): HIV-1 Tat protein stimulates transcription by binding to a U-rich bulge in the stem of the TAR RNA structure. *EMBO J* 9:4145–4153

Ensoli B, Barillari G, Salahuddin SZ, Gallo RC, Wong-Staal F (1990): Tat protein of HIV-1 stimulates growth of cells derived from Kaposi's sarcoma lesions of AIDS patients. *Nature (London)* 345:84–86

Feinberg MB, Baltimore D, Frankel AD (1991): The role of Tat in the human immunodeficiency virus life cycle indicates a primary effect on transcriptional elongation. *Proc Natl Acad Sci USA* 88:4045–4049

Felber BK, Pavlakis GN (1988): A quantitative bioassay for HIV-1 based on transactivation. *Science* 239:973–947

Felber BK, Hadzoponlou-Cladaras M, Cladaras C, Copeland T, Pavlakis GN (1989): Rev protein of HIV-1 affects the stability and transport of the viral mRNA. *Proc Natl Acad Sci USA* 86:1495–1499

Feng S, Holland EC (1988): HIV-1 tat transactivation requires the loop sequence within TAR. *Nature (London)* 334:165–167

Flanagan WM, Corthesy B, Bram RJ, Crabtree GR (1991): Nuclear association of a T-cell transcription factor blocked by FK-506 and cyclosporin A. *Nature (London)* 352:803–806

Gatignol A, Kuman A, Rabson A, Jeang KT (1989): Identification of cellular proteins that bind to the human immunodeficiency virus type 1 trans-activation-responsive TAR element. *Proc Natl Acad Sci USA* 86:7828–7832

Gatignol A, Buckler WA, Berkhout B, Jeang KT (1991): Characterization of a human TAR RNA-binding protein that activates the HIV-1 LTR. *Science* 251:1597–1600

Gaynor R, Soultanakis M, Kuwabara M, Garcia J, Sigman DS (1989): Specific binding of a HeLa cell nuclear protein to RNA sequences in the human immunodeficiency virus transactivating region. *Proc Natl Acad Sci USA* 86:4858–4862

Gelman IH, Silverstein R (1985): Identification of immediate early genes from herpes simplex virus that transactivate the virus thymidine kinase gene. *Proc Natl Acad Sci USA* 82:5265–5269

Green MR, Treisman R, Maniatis R (1983): Transcriptional activation of cloned human β-globin genes by viral immediate–early gene products. *Cell (Cambridge Ma)* 35:137–148

Hammarskjold ML, Heimer J, Hammarskjold B, Sangwan I, Albert L, Rekosh D (1989): Regulation of HIV *env* expression by the *rev* gene product. *J Virol* 63:1959–1966

Hauber J, Cullen BR (1988): Mutational analysis of the transactivation-responsive region of the human immunodeficiency virus type 1 long terminal repeat. *J Virol* 62:673–679

Ho DD, Moudgil T, Alam M (1989): Quantitation of human immunodeficiency virus type 1 in the blood of infected persons. *N Engl J Med* 321:1621–1625

Hoxie JA, Haggarty BS, Rackowski JL (1985): Persistent noncytopathic infection of normal human T lymphocytes with AIDS-associated retrovirus. *Science* 229:1400–1402

Hoxie JA, Alpers JD, Rackowski JL, Huebner K, Haggarty BS, Cedarbaum AJ, Reed JC (1986): Alterations in T4 (CD4) protein and mRNA synthesis in cells infected with HIV. *Science* 234:1123–1127

Hsu MC, Schutt AD, Holly M, Slice LW, Sherman MI, Richman DD, Potash MJ, Volsky DJ (1991): Inhibition of HIV replication in acute and chronic infections *in vitro* by a Tat antagonist. *Science* 254:1799–1802

Hsu MC, Dhingra U, Earley JV, Holly M, Keith D, Nalin C, Richou AR, Schutt AD, Tam SY, Potash MJ, Volsky DJ, Richman DD (1993): Inhibition of HIV-1 replication by Tat antagonists to which the virus remains sensitive after prolonged exposure in vitro. *Proc Natl Acad Sci USA*, in press

Jabbar MA, Nayak DP (1990): Intracellular interaction of human immunodeficiency virus type 1 (ARV-2) envelope glycoprotein gp160 with CD4 blocks the movement and maturation of CD4 to the plasma membrane. *J Virol* 64:6297–6304

Jakobovits A, Smith DH, Jakobovits EB, Capon DJ (1988): A discrete element 3′ of human immunodeficiency virus 1 (HIV-1) and HIV-2 mRNA initiation sites mediates transcriptional activation by an HIV transactivator. *Mol Cell Biol* 8:2555–2561

Jonat C, Rahmsdorf HJ, Park K-K, Cato ACB, Gebel S, Ponta H, Herrlich P (1990): Antitumor promotion and antiinflammation: Down-modulation of AP-1 (Fos/Jun) activity by glucocorticoid hormone. *Cell (Cambridge Ma)* 62:1189–1204

Kao SY, Calman AF, Luciw PA, Peterlin BM (1987): Anti-termination of transcription within the long terminal repeat of HIV-1 by tat gene product. *Nature (London)* 330:489–493

Kessler M, Mathews MB (1991): Tat transactivation of the human immunodeficiency virus type 1 promoter is influenced by basal promoter activity and the simian virus 40 origin of DNA replication. *Proc Natl Acad Sci USA* 88:10018–10022

Larder BA, Darby G, Richman DD (1989): HIV with reduced sensitivity to zidovudine (AZT) isolated during prolonged therapy. *Science* 243:1731–1734

Laspia MF, Rice AP, Mathews MB (1989): HIV-1 Tat protein increases transcriptional initiation and stabilizes elongation. *Cell (Cambridge Ma)* 59:283–292

Malim MH, Hauber J, Fenrick R, Cullen BR (1989): Modulation of the gene expression of regulatory genes of the human immunodeficiency virus by its *rev* gene. *Nature (London)* 335:181–183

Marciniak RA, Calnan BJ, Frankel AD, Sharp PA (1990): HIV-1 Tat protein transactivates transcription *in vitro*. *Cell (Cambridge Ma)* 63:791–802

McDougal JS, Mawle A, Cort SP, Nicholson JKA, Cross GD, Scheppler-Campbell JA, Hicks D, Sligh J (1985): Cellular tropism of the human retrovirus HTLV-III/LAV: I. Role of T cell activation and expression of the T4 antigen. *J Immunol* 135:3151–3162

Nabel G, Baltimore P (1987): An inducible transcription factor activates expression of human immunodeficiency virus in T cells. *Nature (London)* 326:711–713

Nevins JR (1981): Mechanism of activation of early viral transcription by the adenovirus E1A gene product. *Cell (Cambridge, MA)* 26:213–220

Nunberg JH, Schlief WA, Boots EJ, O'Brien JA, Quintero JC, Hoffman JM, Enini JM, Goldman ME (1991): Viral resistance to human immunodeficiency virus type 1-specific pyridinone reverse transcriptase inhibitors. *J Virol* 65:4887–4892

Rice AP, Mathews MB (1988): Transcriptional but not translational regulation of HIV-1 by the *tat* gene product. *Nature (London)* 332:551–553

Richman D, Shih CK, Lowy I, Rose J, Prodanovich P, Goff S, Griffin J (1991): Human immunodeficiency virus type 1 mutants resistant to nonnucleoside inhibitors of reverse transcriptase arise in tissue culture. *Proc Natl Acad Sci USA* 88:11241–11245

Roy S, Delling U, Chen CH, Rosen CA, Sonenberg N (1990): A bulge structure in HIV-1 TAR RNA is required for Tat binding and Tat-mediated transactivation. *Genes Dev* 4:1365–1373

Schnittman SM, Greenhouse JJ, Psallidopoulos MC, Baseler M, Salzman NP, Fouce AS, Lane HC (1990): Increasing viral burden in CD4+ T cells from patients with human immunodeficiency virus (HIV) infection reflects rapidly progressive immunosuppression and clinical disease. *Ann Intern Med* 113:438–443

Selby MJ, Peterlin BM (1990): Transactivation by HIV-1 Tat via a heterologous RNA binding protein. *Cell (Cambridge, MA)* 62:769–776

Selby MJ, Bain ES, Luciev PA, Peterlin BM (1989): Structure, sequence, and position of the stem-loop in TAR determine transcriptional elongation by Tat through the HIV-1 long terminal repeat. *Genes Dev* 3:547–558

Shahabuddin M, Volsky B, Hsu MC, Volsky DJ (1992): Restoration of cell surface CD4 expression in human immunodeficiency virus type 1–infected cells by treatment with a Tat antagonist. *J Virol* 66: 6802–6805

Sheline CT, Milocco LH, Jones KA (1991): Two distinct nuclear transcription factors recognize loop and bulge residues of the HIV-1 TAR RNA hairpin. *Genes Dev* 5:2508–2520

Siliciano RF, Lawton T, Knall C, Karr RW, Berman P, Gregory T, Reinherz EL (1988): Analysis of host–virus interactions in AIDS with anti-gp120 T cell clones: Effect of HIV sequence variation and a mechanism for CD4+ cell depletion. *Cell (Cambridge MA)* 54:561–575

Sodroski JG, Rosen CA, Wong-Staal F, Salahuddin SZ, Popovic M, Arya S,

Gallo RC (1985): Trans-acting transcriptional regulation of human T-cell leukemia virus type III long terminal repeat. *Science* 227:171–173

Stenberg RM, Depto AS, Fortney J, Nelson JA (1989): Regulated expression of early and late RNAs and proteins from the human cytomegalovirus immediate–early gene region. *J Virol* 63:2699–2708

Tanneau F, McChesney M, Lopez O, Sansonetti P, Montagnier L, Riviere Y (1990): Primary cytotoxicity against the envelope glycoprotein of human immunodeficiency virus-1: Evidence for antibody-dependent cellular cytotoxicity *in vivo*. *J Infect Dis* 162:837–844

Viscidi RP, Mayur K, Lederman HM, Frankel AD (1989): Inhibition of antigen-induced lymphocyte proliferation by Tat protein from HIV-1. *Science* 246:1606–1608

Volsky DJ, Pellegrino MG, Li G, Logan KA, Aswell JE, Lawrence NP, Decker SR (1990): Titration of human immunodeficiency virus type 1 (HIV-1) and quantitative analysis of virus expression *in vitro* using liquid RNA-RNA hybridization. *J Virol Methods* 28:257–272

Weeks KM, Ampe C, Schultz SC, Steitz TA, Crothers DM (1990): Fragments of the HIV-1 Tat protein specifically bind TAR RNA. *Science* 249:1281–1285

Wu F, Garcia J, Sigman D, Gaynor R (1991): Tat regulates binding of the human immunodeficiency virus transactivating region RNA loop-binding protein, TRP-185. *Genes Dev* 5:2128–2140

8

Discovery of Pirodavir, a Broad-Spectrum Inhibitor of Rhinoviruses

KOEN ANDRIES

Introduction

For many years there has been an intensive search for antivirals that are active against rhinoviruses. The reason is very simple: rhinoviruses are the main culprits of the common cold, and although admittedly the common cold is a very mild affliction, almost all of us from time to time use medicine to relieve our cold symptoms. Given the fact that a specific antiviral agent is not included in any of the drugs currently marketed for the treatment of colds, the potential market for an effective common cold drug is huge.

The benign nature of the disease is certainly not an indication of the difficulty of the field. There are five formidable obstacles in the development of a drug against the common cold. All five obstacles will need to be addressed before a candidate drug can reach the market.

The first obstacle is the multitude of viruses that cause colds. The presence of exactly 100 serotypes of rhinoviruses is, on the one hand, a reason why the possibility of a vaccine is very remote, but, on the other hand, it poses a formidable challenge to those who want to develop an antiviral that is active against as many of these serotypes as possible. In addition, although perhaps somewhat more than 50% of colds are caused by rhinoviruses (the use of new techniques in the diagnosis will tell us soon), the other 50% are probably not. This creates a significant marketing problem if a drug that is solely active against rhinoviruses is to be developed. It is possible that combinations with existing drugs providing symptomatic relief will partially solve the problem, but combinations of drugs are not very easy to develop either.

The second obstacle arises from the fact that the common cold is a

The Search for Antiviral Drugs
Julian Adams and Vincent J. Merluzzi, Editors
© 1993 Birkhäuser Boston

nasal infection. The logical approach is to bring the antiviral to the place where the virus replicates and thus use a nasal spray, especially if one possibly wants to combine the antiviral with compounds providing symptomatic relief. Another reason to develop an intranasal formulation for the drug is that even the slightest of side effects will not be acceptable for a drug that treats the common cold. However, it is very difficult to ensure that adequate drug levels are achieved and persist in the nasal environment. Mucociliary transport and absorption reduce the concentration of almost any chemical that is administered into the nose by half within just 15–20 minutes. This probably explains why we have to administer so many antiviral units to achieve clinical efficacy (see below).

The third hurdle is the difficulty in proving the clinical efficacy of a drug. The absence of an animal model dictates that any candidate be developed straight from the test tube to tolerance and efficacy studies in human volunteers. The implication is that, of course, a full IND (U.S. Investigative New Drug Application) or CTX (U.K. Clinical Trial Exemption) file must be submitted with adequate toxicity data before one can even think of starting a clinical trial. The second difficulty here is the benign nature and short duration of the experimental infection in humans. These make it very difficult to prove the existence of anything less than a dramatic effect of the drug.

The fourth obstacle is our complete absence of understanding of the pathogenesis of the rhinovirus infection. For instance, we do not know to what extent the symptoms of a cold are caused by viral replication that is still continuing at that time. In the worst case, all symptoms are caused as a direct consequence of viral replication that occurs during the incubation period. It is, of course, tempting to speculate that this is the reason why a therapeutic effect in an experimental trial has yet to be found. The problem is that we will only know whether a therapeutic effect can be achieved when we actually see it. In the meantime, we can only try as hard as we can to reach the ultimate goal and hope for the best. After all, it is only a few decades since everybody speculated that selective inhibitors of viral replication would never be found!

The fifth and last obstacle is the possibility of rapid emergence of resistance to a potential antirhinoviral compound. This is not a problem specific to the rhinovirus field, but one that generally applies to all inhibitors of viruses and bacteria. Premature interest in resistance can, however, prevent the further development of a project before it has really started.

Start of Antiviral Drug Research

When I joined the Janssen Research Foundation (JRF) in Beerse, Belgium, at the end of 1982, a small antiviral program had already been initiated about a year earlier. Dr. Paul Janssen, president of the JRF, had become interested in antivirals for herpesviruses and rhinoviruses. One of my colleagues (Dr. Marc De Brabander) went to the Common Cold Unit in England in 1981. Dr. David Tyrrell, who has always been at the forefront of clinical research on rhinoviral infections, demonstrated the productive challenge experiments he had established in human volunteers with HRV9 (human rhinovirus serotype 9) and HRV2, and kindly provided us with both these viruses and the cells. Back home in Beerse there was no problem in growing the viruses, and thus a test system was designed.

Test Systems

The use of a viral infectivity test was preferred over the possible use of individual viral enzymes or proteins (e.g., viral protease). Infectivity tests are designed to allow for several viral replicative cycles before the development of a complete cytopathic effect (CPE), thus ensuring that an antiviral compound active at any of the possible steps of the replication will be detected. A compound that is active against an individual viral enzyme or protein may also be detected in an infectivity assay, but the reverse is not always true. In addition to that, isolated targets for antirhinoviral compounds had not yet been identified.

Both HRV9 and HRV2 produced a distinct CPE in both the HeLa cell line that we had obtained from the Common Cold Unit and in human fibroblasts. The former were of course more convenient to use. Microscopic evaluation of microtiter plates was used for the assessment of the antiviral activity in the early years. Typically, four drugs were tested per microtiter plate in fivefold dilutions starting from 250 μg/mL and down to 0.4 μg/mL. Cytotoxicity controls were included in each test and for each concentration tested. After an incubation of 4–5 days at the recommended temperature (33 °C, that of the human nasopharynx), the plates were read and the concentrations inhibiting the viral CPE by 0–25% and 75–100% were recorded. Similarly, the concentrations that proved to be cytotoxic for 0–25% and 75–100% of the cells were scored. This gave us a very crude estimation of a drug's antiviral and cytotoxic properties.

This first version of our antiviral test was subsequently replaced by a more sophisticated one. Basically, the differences were that twofold instead of fivefold dilutions were used, that the drug was allowed to incubate with the virus for 1 hour before they were both added to subconfluent HeLa cells, and that the 0%, 25%, 50%, 75%, and 100% percentiles of CPE reduction were scored at the end of the incubation period. This procedure allowed for more quantitative estimates of viral cytopathogenicity and for the calculation of the 50% inhibitory concentration (IC_{50}) for a particular serotype (Andries et al., 1988).

The second method, however, still had a few drawbacks. All dilutions were still made manually, and for practical reasons we did not change the micropipet tips between the dilutions. That did not raise serious problems until very potent and at the same time very hydrophobic compounds were found. Because of their hydrophobicity, these antiviral compounds adhere strongly to plastic tips when serially diluted in a watery medium. HPLC analysis showed that when a serial dilution program does not include a tip change with a frequency of every three fivefold dilutions, compound that is adsorbed to the plastic tips at the highest concentrations is washed off the tip in the lowest concentrations. The absence of pipet changes in the second method sometimes seriously exaggerated the IC_{50} values of the most potent compounds. Another drawback of the second method was that the IC_{50} values were still a rather rough estimation and that the cytotoxicities were evaluated on subconfluent cell monolayers.

One of my collaborators (Rudy Willebrords) recently designed a fully automated procedure, using a robotic system that operates in an incubation room and allows for plastic tip changes whenever required. Suspensions of HeLa cells are added to the microtiter plates after dilution of the antiviral compounds, rather than the virus–drug mixtures being inoculated onto subconfluent monolayers. The viability of mock- and virus-infected cells is quantitated spectrophotometrically by a tetrazolium colorimetric method, the MTT [3-(4,5-dimethylthiazol-2-yl)-2,5-diphenyl-2H-tetrazolium bromide] assay (Andries et al., 1992). Similarly, IC_{50} and CC_{50} values can now be calculated to quantitate the antiviral potency and the 50% cytotoxic concentration of the same compound. From the IC_{50} and the CC_{50} values, a selectivity index, or SI, can be calculated for each individual serotype. Compounds with high SIs may be less toxic for use in humans, although various side effects such as organ specific toxicity, taste alterations after intranasal use, or nasal irritation can of course not be predicted from the in vitro results.

At the start of the antiviral program, we were just randomly screening compounds. After a while, it was decided that a selection of 600 compounds would be made from the JRF compound repository, all prototypes of different chemical series, and all of them being pharmacologically inert. It was also decided to synthesize a few analogues of 4',6-dichloroflavan, the most potent antirhinoviral compound described at that time (Bauer et al., 1981). About 20 compounds were screened weekly against just two viruses: HRV9 and herpes simplex virus type 2. Each screening was performed by one of two technicians who spent about half of their time doing the tests and taking care of the cell cultures. In 1983, the two technicians were diverted to other projects and were replaced by one full-time technician.

The First Leads

During the first year, several dozens of compounds with moderate activity in our test system were identified. Most of them were to some extent cytotoxic as well, but we were puzzled by the fact that some compounds displayed antiviral activity at those borderline concentrations, whereas others did not. In the end, we decided to select only those compounds that were active at concentrations not cytotoxic for the cells. All resulting compounds were, with one exception, analogues of 4',6-dichloroflavan. The IC_{50}'s of the most potent 4',6-dichloroflavan analogues were around 2 μg/mL. However, we became more and more impressed by the results obtained with one particular compound, coded R 51174 (Figure 1) (the R standing for research number), which was unrelated to 4',6-dichloroflavan. The compound had been tested three times, each time resulting in a different IC_{50} value (2, 50, and 10 μg/mL, respectively), probably because at that time we did not take too much care about using a standard viral dose in each test. As we could not find cytotoxicity at a concentration of 250 μg/mL, the antiviral activity appeared to be specific. When we ran the compound in parallel with the 4',6-dichloroflavan analogues against HRV9, we found that R 51174 had a potency that was only five times lower than that of 4',6-dichloroflavan.

Close analogues to R 51174 were not to be found in the JRF depository. Actually, it was pure coincidence that the compound received a research code number in the first place. In normal circumstances, a compound research number is only granted if the compound is chemically new. In this case, the chemist had synthesized more than he could use as a building block for the synthesis of ketoconazole analogues, and a code number was given, although the *ortho*-methoxy analogue of R 51711 had been described in the chemical literature.

We repeatedly showed our results to Dr. Paul Janssen, who visited the laboratory very regularly, to Dr. Marcel Janssen, the head of the chemistry department, and to Dr. Karel Schellekens, the coordinator of research. The compound had an interesting potency without any synthesis based on structure–activity relationship (SAR) having been undertaken. We also confirmed that the compound had specific antiviral activity against HRV2. After a few weeks, Dr. Paul Janssen decided to initiate a new synthesis program around the lead molecule.

As R 51174 is a relatively simple structure to synthesize, the two chemists that were assigned to this project (Dr. Raymond Stokbroekx and Mr. Johan Willems) soon succeeded in delivering the first two analogues. One of these, bearing a trifluoromethyl substituent in the meta-position of the benzene ring, had an IC_{50} slightly better than R 51174. The next series of 11 compounds became available in the last week before the summer holiday break of 1983. We were so keen to

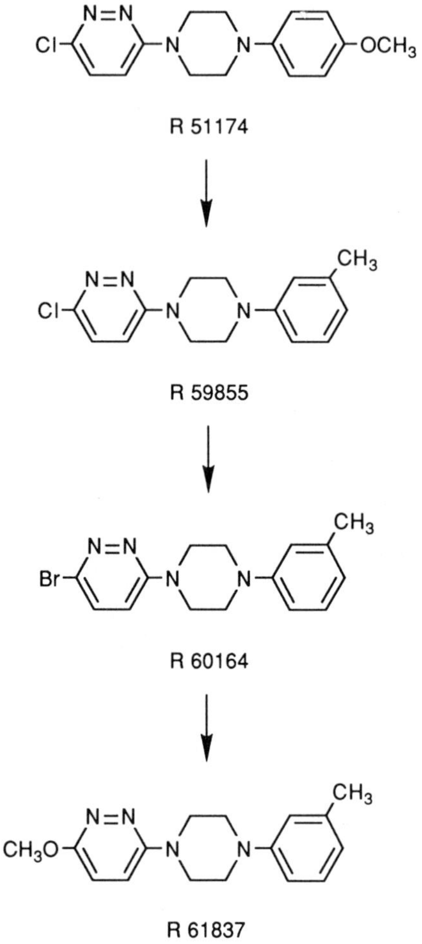

R 51174

R 59855

R 60164

R 61837

FIGURE 1. The first pyridazinamines.

know whether we were on the right track that we kept the cells running during the break so that a test could be performed immediately after the holiday. The results were almost too good to believe. Among this series of 11 analogues, 8 were active, 5 of them at lower concentrations than R 51174 itself. One week later, we identified a compound (R 59855, Figure 1) with an IC_{50} 25 times lower than that of R 51174. The cytotoxicity of the compounds was always above 50 μg/mL. Needless to say, everybody involved got very excited.

Mechanism of Action

In June 1983, Dr. Margaret Tisdale from Burroughs Wellcome presented a poster on the mechanism of action of 4',6-dichloroflavan at the NATO Advanced Study Institute meeting in Les Arcs, France. She concluded that 4′,6-dichloroflavan was a capsid-binding compound, based on the fact that the compound inactivated susceptible serotypes and that in density gradient experiments radiolabeled compound peaked in the same fractions as the virions themselves. I also learned about the use of a time-of-addition experiment to determine at what stage of the viral replication cycle a compound exerts its antiviral activity. Back home in Beerse we set up a similar time-of-addition experiment with R 60164. The results showed convincingly that the compound acted at a very early stage of the viral replication, coincident with the susceptible stage of 4′,6-dichloroflavan, and clearly earlier than the stage susceptible to enviroxime (Figure 2). Enviroxime, or *E*-2-amino-1-(isopropyl sulfonyl)-6-benzimidazole phenyl ketone oxime, was known to inhibit the initiation of RNA synthesis of rhinoviruses (Anonymous, 1984).

During the following months, it became clear that our compounds stabilized susceptible rhinoviruses against degradation by acid and heat, indicating that they too were capsid binders (Andries et al., 1988) (Figure 3). Experiments using light-sensitive virus established that one

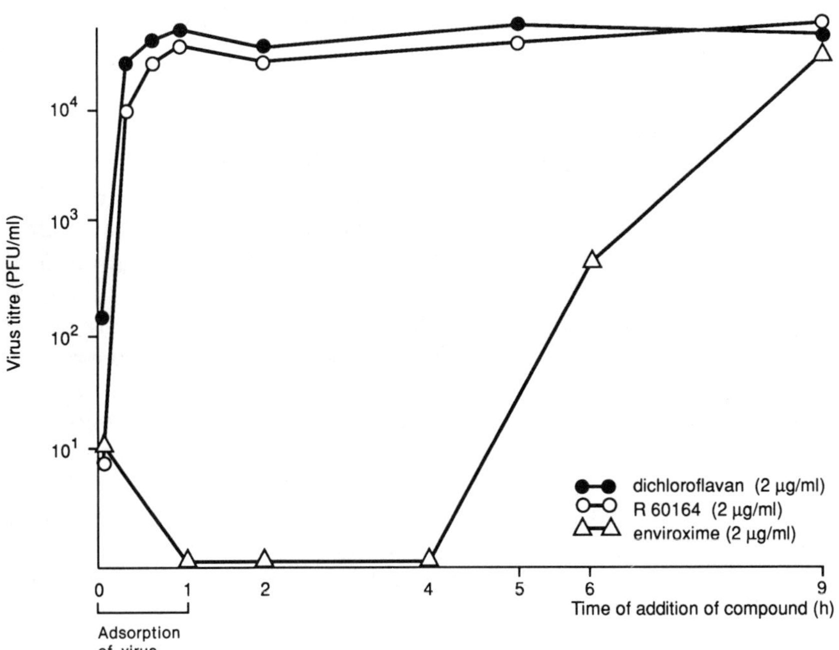

FIGURE 2. Effect of time of addition of pyridazinamines to rhinovirus-infected cultures. Cells were infected at a multiplicity >1 with HRV9. After 60-min incubation at 33 °C, unadsorbed virus was removed by three washing steps. Dichloroflavan (2 µg/mL), R 60164 (2 µg/mL), or enviroxime (2 µg/mL), an inhibitor of the initiation of RNA replication, was added at the times indicated during the virus growth cycle. At 9 hours, the cells plus medium of all cultures were freeze-thawed three times and the virus yields were determined by plaque assay.

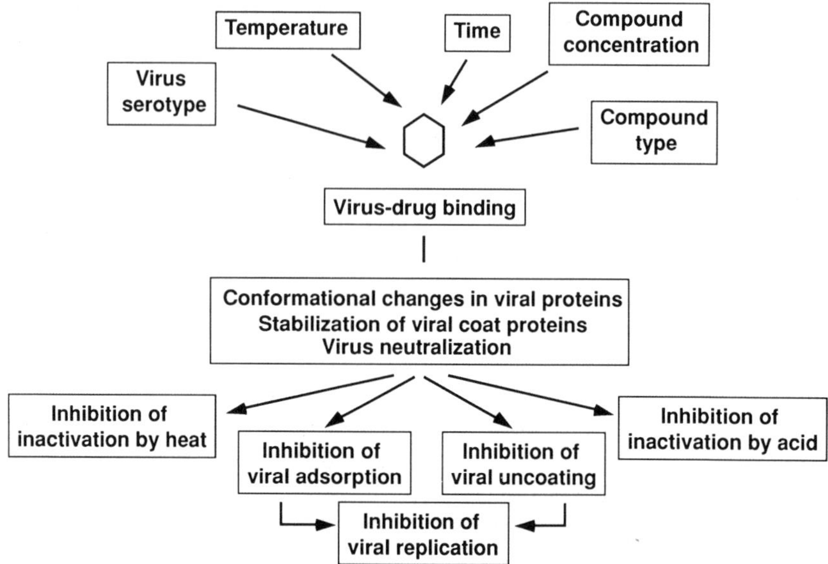

FIGURE 3. Overview of the factors involved in the binding of pyridazinamines to rhinoviruses.

of the compounds had an effect on the uncoating of HRV9. Although our compounds were called uncoating inhibitors by others, we were careful not to make firm statements on their mechanism of action. We and others have only recently detected that capsid binders have additional effects on the adsorption and maybe also on the penetration of some but not all serotypes. The extent of the binding of drug to the viral particles is influenced by the incubation temperature, the nature of the compound and its concentration, and the virus serotype. Binding results in neutralization of the virus infectivity. In contrast to the results obtained with the other capsid-binding compounds, the neutralization obtained with pyridazinamines can sometimes not be reversed by extraction of the compound with organic solvents (Andries et al., 1988). Interestingly, we also established that there was no direct (quantitative) correlation between the stabilization event and the inhibition of replication (Andries et al., 1989). Both effects are apparently independent consequences of the binding of the drug to the virus.

Choosing a Development Candidate and a Suitable Formulation

In the autumn of 1983 and in 1984, we had a few informal meetings with members of the Common Cold Unit (Dr. Bob Phillpotts, Dr. David

Tyrrell) and of the Janssen U.K. company (Dr. Natalie Taylor), in both Beerse and Salisbury. Just before the first of these visits, we prepared a brief dossier to convince our visitors about the antiviral potency of the compounds. Our guests seemed impressed by the potency that R 59855 exerted in plaque reduction assays, using HRV2 and HRV9.

Dr. Tyrrell, director of the unit, stressed the need to perform additional tests in human nasal epithelial cells, because some compounds appeared to be cell specific. He also proposed testing the compound of choice against a few wild-type strains and a few challenge virus strains that he would provide to us. He felt that testing in humans would only be warranted if a member of the series could be found with a potency at least 10 times better than enviroxime, as the power of detection of the trials is low owing to the small number of volunteers and low attack rates. Last but not least, he stressed the need to use as many antiviral units (defined as the IC_{50}) as possible in the formulation to be given to the volunteers. None of the synthetic compounds tested in humans thus far, including 4',6-dichloroflavan, had achieved a clinical effect. The only positive trials were those in which interferon was used at fairly high concentrations (at least 1×10^6 antiviral units).

The experiment using human nasal epithelial cells was not difficult to perform. Human nasal polyps were obtained from patients undergoing polypectomy, and fragments of these were cultured and infected with HRV9 previously incubated with various concentrations of the compound. The reductions of viral yields obtained in human epithelial cells were roughly comparable to those obtained in HeLa cells, although there was a slight tendency toward less inhibition in the former.

We performed SAR studies for somewhat more than a year. Although it was felt that the spectrum of activity against the various serotypes was important, in the first instance a potent activity against one challenge virus would give an indication as to whether this series of compounds had antiviral activity in humans or not. Dr. Paul Janssen finally selected R 61837 (Figure 1) as the molecule of choice. The compound's IC_{50} was 0.006 μg/mL for HRV9, and it had a CC_{50} of 15 μg/mL.

The formulation was the most difficult issue to tackle. The antiviral pyridazinamines are known to bind to hydrophobic amino acids of the viral capsid (see below). Hydrophobicity of the compounds is thus essential to achieve high antiviral potency. On the other hand, compounds that are not in solution are not very likely to be active. They probably need to be applied in an aqueous solution to enable the interception of viral particles in the aqueous perciliary layer underlying the mucous blanket on the nasal mucosa. A solution containing a high concentration of a highly hydrophobic compound was thus needed. As the maximal solubility of our molecules was usually in the range of 1–20 μg/mL and the antiviral potency in the range of 5–10 ng/mL, the solubility had to be improved by a factor of 1000 to achieve approximately 10^6 antiviral units per milliliter of the formulation. This very difficult technical problem was, in the end, solved by Dr. Jean Mesens, our colleague at the JRF Pharmaceutical Development Department. He proposed to try cyclodextrin derivatives. Cyclodextrins are cyclic polysaccharides with an interior hydrophobic cavity and an exterior hydrophilic surface (Figure 4). The antiviral molecule inserts itself into the hydrophobic cavity, and the complex can be held in solution under certain physicochemical conditions. At first, we were worried that R 61837, once inserted in the cyclodextrin molecule, would no longer be able to exert its antiviral activity, but a few experiments proved that that was not a problem. Several cyclodextrin derivatives were tested before hydroxypropyl-β-cyclodextrin was identified as a molecule with the desired properties (nonirritating, nontoxic). By its use, the solubility of several pyridazinamines could be increased by a factor of 1000. A nasal spray containing 10% hydroxypropyl-β-cyclodextrin and 2.5 mg/mL of the pyridazinamine R 61837 was subsequently developed.

In the autumn of 1985, all problems had been resolved and Dr. Janssen decided to initiate all studies needed for the submission of a CTX for R 61837 in the United Kingdom.

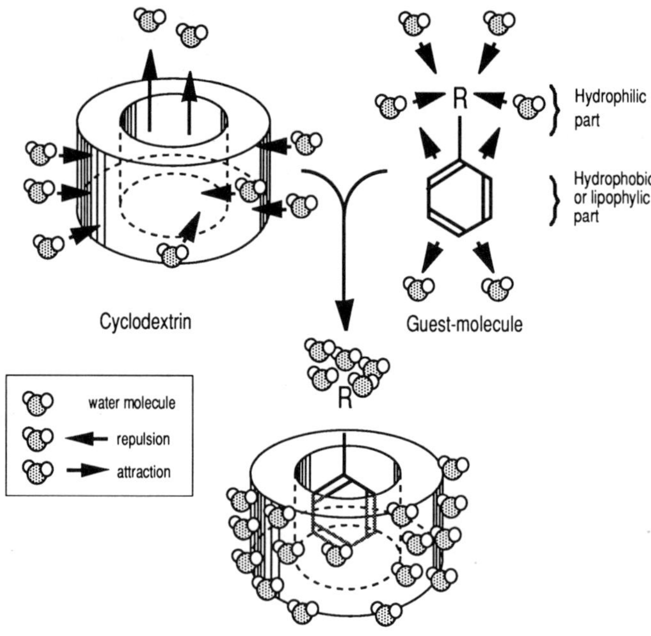

FIGURE 4. Cyclodextrins are hydrophilic molecules with an interior hydrophobic channel that can be used to solubilize highly hydrophobic compounds.

First Clinical Trials at the Common Cold Unit

The most important goal of the human trials was for us to learn whether capsid-binding compounds were active in humans or not. We decided to give the drug the maximal chances we could. Thus a prophylactic trial was designed with a total of 25 doses (6 doses of 0.2 mL of drug or placebo in each nostril per day), given commencing 24 hours before the virus challenge and continuing for a further 3 days (Al-Nakib et al., 1989). The treatment scheme of 6 doses per nostril per day (each dose consisting of two puffs of 0.1 mL each) was considered to be demanding but possible. The study would be preceded by a limited tolerance study of intranasal administration.

We decided to use HRV9 as the challenge virus because of its high sensitivity to R 61837 and the experience that had been built up in the Common Cold Unit with that serotype. The high potency of R 61837 against the challenge virus, combined with the use of hydroxypropyl-β-cyclodextrin to achieve high concentrations of R 61837 in an acqueous solution, made it possible for the first time to use an amount of antiviral units comparable to those used during successful interferon trials.

The first trial, which required three intakes of volunteers in the summer of 1986, was a real thriller. After its completion, we received a first message saying that the result was negative. Indeed, there were as many colds in the treated group (6 out of 28 challenged) as there were in the placebo group (6 out of 27 challenged). We were severely depressed, and we already saw the antiviral program being turned down. However, after a few bad weeks the good news came through. It appeared that no colds had occurred in the drug group while medication was given but that they had all appeared only after medication was ceased (Figure 5, panel A). This of course changed everything. It was clear that the drug had worked, and all investigators involved were very pleased by the result. It was the first time that a synthetic antirhinovirus compound had produced a clinical effect in humans.

In a meeting held in Beerse at the end of 1986, we all agreed that the drug had not been given for a long enough period. At the same time, the fact that the same percentage of volunteers in both groups had seroconverted proved that the drug had not prevented the infection but really did interfere with the viral replication. Thus, the start of treatment 24 hours before the virus challenge was felt to be unnecessary.

Dr. Widad Al-Nakib of the Common Cold Unit thus designed a second prophylactic trial in which treatment was given for a longer period, starting just 4 hours before challenge. During the 7 days of observation after the challenge, symptoms and in this case also the number of colds (two against seven) were greatly reduced in the drug group (Figure 5, panel B).

In a third trial, we wanted to find out what would happen if drug

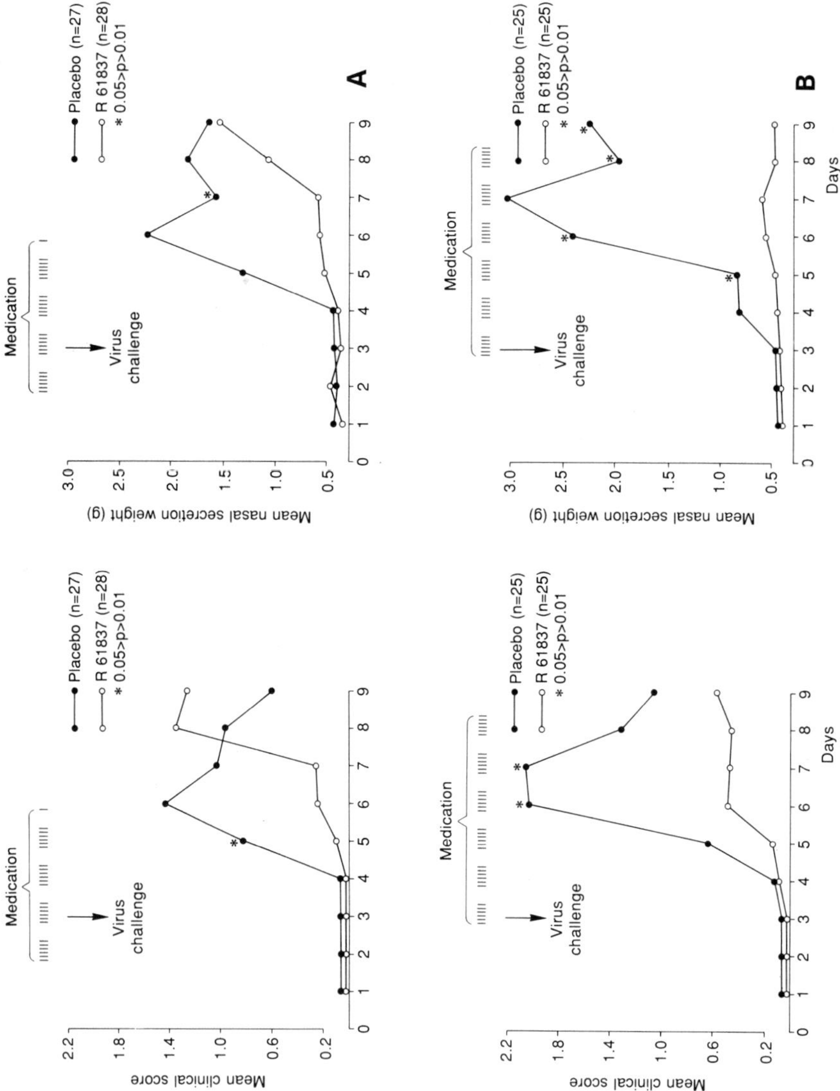

Placebo (n=27)
R 61837 (n=28)
* 0.05>p>0.01
A
Medication
Virus challenge
Mean nasal secretion weight (g)
3.0
2.5
2.0
1.5
1.0
0.5
0 1 2 3 4 5 6 7 8 9
Placebo (n=25)
R 61837 (n=25)
* 0.05>p>0.01
B
Medication
Virus challenge
Mean nasal secretion weight (g)
3.0
2.5
2.0
1.5
1.0
0.5
Days
0 1 2 3 4 5 6 7 8 9
Placebo (n=27)
R 61837 (n=28)
* 0.05>p>0.01
Medication
Virus challenge
Mean clinical score
2.2
1.8
1.4
1.0
0.6
0.2
0 1 2 3 4 5 6 7 8 9
Placebo (n=25)
R 61837 (n=25)
* 0.05>p>0.01
Medication
Virus challenge
Mean clinical score
2.2
1.8
1.4
1.0
0.6
0.2
Days
0 1 2 3 4 5 6 7 8 9

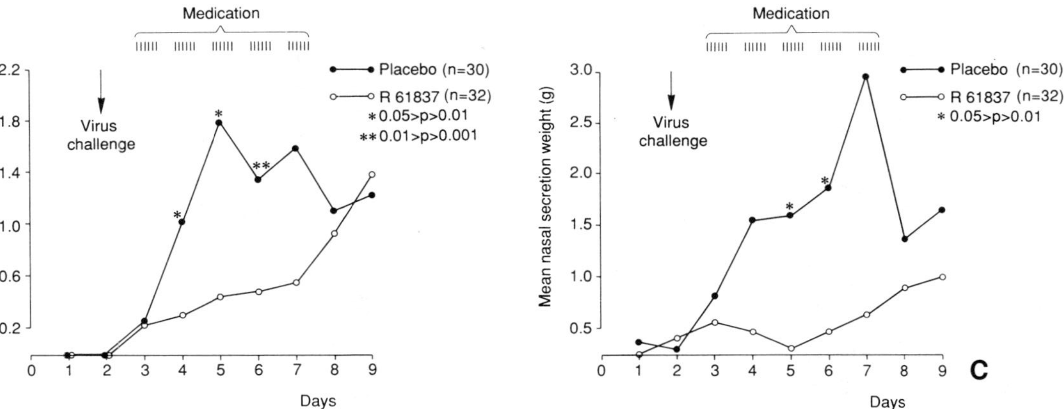

FIGURE 5. Overview of clinical trials in human volunteers with R 61837. Mean daily clinical scores and nasal secretion weights in trials A, B, and C, in top, middle, and bottom panels, respectively. Reprinted with permission of Blackwell Scientific Publications, Ltd., from Barrow GI et al. (1990): An appraisal of the efficacy of the antiviral R 61837 in rhinovirus infections in human volunteers. *Antiviral Chem Chemother* 1:279–285.

treatment were to be started after the virus challenge was given, i.e., in the incubation period. In this semitherapeutic trial a total of 30 doses was given, with medication starting 22 hours after challenge. Again, both symptoms and colds were greatly reduced in the drug group. Differences in total clinical scores and total nasal secretion weights reached statistical significance ($p < 0.05$) (Figure 5, panel C).

In a final therapeutic trial, dosing was initiated after symptoms had appeared, and administration occurred every waking hour up to a maximum of 15 doses a day until the end of the trial, which lasted 5 days. No beneficial effect on the progression of the colds treated with R 61837 was seen. Indeed, the mean total clinical score was significantly greater for those receiving drug than for those given placebo.

The last study was of course a major disappointment. We felt that the slightly irritating effect of the drug might have masked a possibly beneficial effect on the development of symptoms, but that was something very difficult to elaborate on. In addition, we realized that little or no prophylactic effect may occur when the great majority of other rhinovirus serotypes are the infectious agent, as these are much more resistant to the antiviral effect of R 61837 than the challenge virus HRV9. Further development of compounds with a higher potency and a much broader spectrum was thus needed to increase the prophylactic or therapeutic usefulness of this class of compounds.

Discovery of Two Opposite Spectra of Capsid-Binding Compounds

Shortly after the discovery of the first pyridazinamines, we tried to get hold of reference compounds that were supposed to have the same mechanism of action. We thus acquired samples of the capsid binders 4′,6-dichloroflavan and a chalcone (Ishitsuka et al., 1982). Both compounds had been reported to inhibit the replication of several rhinovirus serotypes, although the IC_{50}'s varied considerably from serotype to serotype. We initiated a study using 40 rhinovirus strains in a plaque titration reduction test to compare the potencies and spectra of two selected pyridazinamines and the two reference compounds.

The design of this test is very simple and allows several compounds to be tested against new virus strains without us having to bother too much about the titer of the virus stock. Tenfold dilutions of the virus stock are inoculated into 6-well Petri dishes, and after an adsorption period of 1 hour maintenance medium with a fixed concentration (2 μg/mL in this case) of the compound to be tested is added in an agarose overlay. The reduction of the virus titer in the presence of the compound gives a rough estimation of the virus susceptibility.

We found that many but not all of the serotypes were susceptible to the compounds and that the extent of the antiviral activity was very variable. Quite unexpectedly, however, we noticed that the reference compounds had spectra (Figure 6) that were very similar to each other and very similar to the spectrum of the pyridazinamines. Although the chalcone was clearly the most potent compound among those tested and thus had the broadest spectrum, most serotypes were either inhibited by all compounds or by none of them. We had hoped that compounds with complementary spectra could be found, so that one could think of developing a mixture of compounds to achieve broad-spectrum activity. We really did not understand that compounds that are so different in structure all exerted an antiviral effect on some serotypes, whereas other serotypes were not susceptible to any of them. When we subsequently used an IC_{50} test and included more reference capsid-binding compounds [e.g., SDS (sodium decyl sulfate) and RMI 15731], we obtained the same picture. It was almost as if there were two kinds of rhinovi-

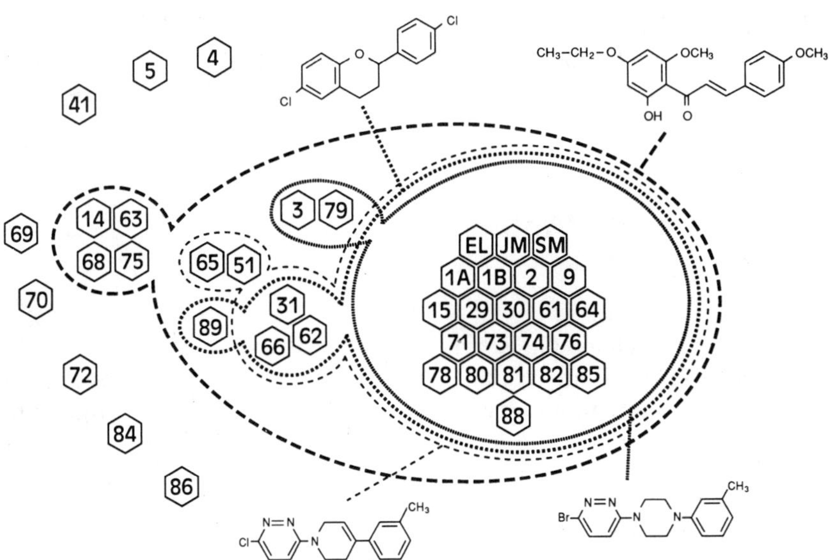

FIGURE 6. Antirhinovirus spectra of chalcone, dichloroflavan, and two pyridazinamines. The susceptibility of 40 rhinovirus serotypes was assessed. Tenfold dilutions of the virus stocks were inoculated into 6-well Petri dishes, and, after an adsorption period of 1 hour, maintenance medium with a fixed concentration (2 μg/mL) of the antiviral compound was added in an agarose overlay. A virus was considered susceptible to a particular compound if the virus titer in its presence was reduced by at least 2 $\log_{10}$. The serotypes that are present within the circle representing a particular compound were susceptible to that compound, whereas those outside the circle were not.

ruses: those susceptible to almost all capsid-binding compounds, and those resistant to all of them.

In 1985, WIN 51711, a capsid-binding compound developed by the Sterling Winthrop group, was described by Otto et al. It took a while before we noticed that this compound had a strange spectrum, in comparison to the consensus spectrum that we had identified. For instance, serotypes 86 and 5 were reported to be most susceptible to WIN 51711, followed by 14 and 3. We had found that these serotypes were most resistant to the other capsid-binding compounds (Figure 6). The spectrum of the WIN compound not only appeared to be different, it almost looked like a mirror image of the spectrum of all other compounds.

As our observation was based entirely on the comparison of IC_{50} results obtained in different laboratories, we first wanted to confirm our suspicions about the existence of opposite spectra by testing the WIN compound in our model, using the same serotypes. When we did, we saw indeed that WIN 51711 had almost no activity against the viruses that were susceptible to the other capsid-binding compounds and that the compound was active against the previously resistant viruses. We were quite excited about this because it meant that the combination of two antivirals to achieve a mixture of compounds with broad-spectrum activity was indeed possible.

Discovery of the Antiviral Binding Site in Rhinoviruses

In 1985, Dr. Michael G. Rossmann and collaborators described the structure of HRV14 in atomic detail. Although this was a paper of outstanding quality in many respects, we were particularly intrigued by the section describing the so-called canyon structure. Dr. Rossmann himself hypothesized that the canyon had to be the region for cell attachment, based on the assumption that receptor binding sites on viruses are likely to be located in surface depressions small and deep enough to be protected from antibodies (Figure 7B). Along the same lines, we reasoned that the canyon had to be the attachment site for capsid-binding compounds. An antiviral binding site on the exterior of the virion was considered to be unlikely, because it would not have been conserved among at least some serotypes, as was apparently the case.

In 1986, several WIN compounds (capsid-binding compounds from the Sterling Winthrop group), including WIN 51711, were shown to bind in a hydrophobic pocket located beneath the canyon structure in HRV14 (Figure 7C) (Smith et al., 1986). These were the first structural studies of antiviral agents complexed with a virion. Because of the

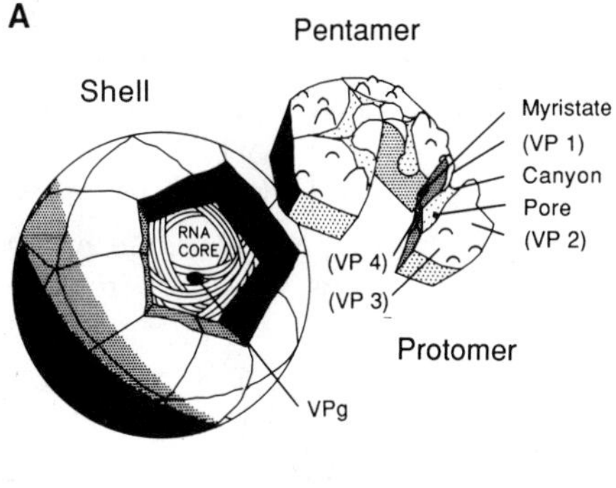

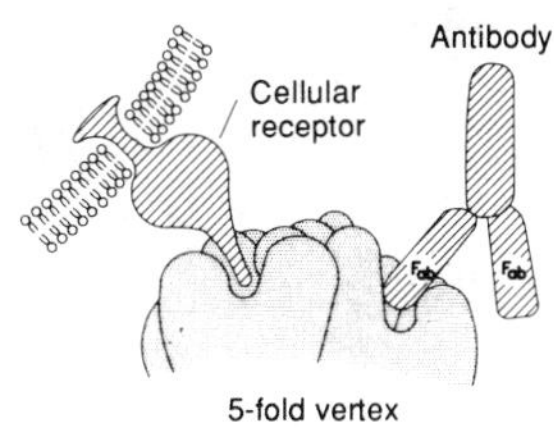

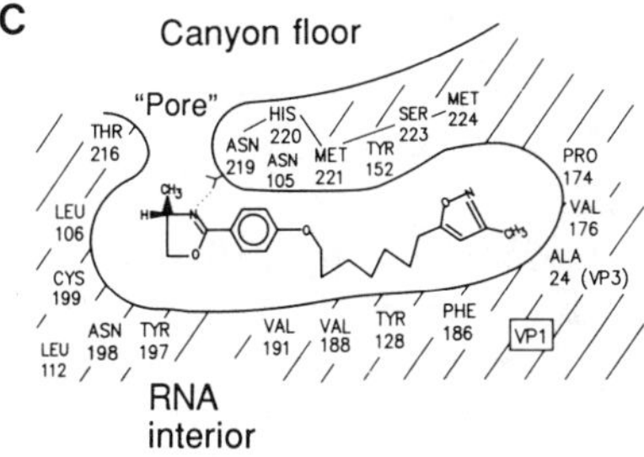

FIGURE 7. Key features in the function of a typical picornavirus. (A) Exploded diagram showing the RNA interior and the location of canyon structure. (B) Binding of cellular receptor to the floor of the canyon. The binding site of the cellular receptor for the majority of rhinoviruses has a diameter roughly half that of an IgG (immunoglobulin gamma) antibody molecule. (C) Location of the drug-binding site in VP1 of HRV14 and identity of amino acid residues lining the wall. The drug depicted here, WIN 52084, prevents attachment of HRV14 by deforming part of the canyon floor. Reprinted with permission of Raven Press from Rueckert RR (1990): Picornaviridae and their replication. In *Virology*, Fields BN, Knipe DM, eds. New York: Raven Press.

existence of two opposite spectra, we considered the possibility that there could be more than one binding site: the one that had been described for the WIN compounds in HRV14, and another one suitable for binding the other capsid-binding compounds.

We decided to raise resistant mutants of some serotypes susceptible to all compounds, although not to the same degree, to see whether we could observe cross-resistance. This was indeed the case. Upon exposure of HRV9 to R 61837 in several passages, the virus became cross-resistant to all other capsid-binding compounds, including the WIN compound, although not to the same extent (Andries et al., 1989). This strongly indicated that there was only one binding site after all, the hydrophobic pocket, not only for WIN 51711 and R 61837 but for all capsid-binding antivirals.

Yet, we were still left with the question of how one binding site could account for the observed differences in spectra. Indeed, how could one explain that, on the one hand, we had a group of antivirals with widely different chemical structures, all apparently binding in the same pocket, and, on the other hand, another chemical had a different spectrum although binding in the same place? We had no idea.

Discovery of Two Separate Rhinovirus Groups Based on Antiviral Sensitivity

We decided to test all 15 capsid-binding antivirals that we knew about at that time against all 100 rhinoviruses and a few enteroviruses in the same test system, to obtain more comparative data on their spectra and to allow for a statistical analysis of the interrelationship between differences in antiviral sensitivities and genetic differences. Indeed, we had the impression that the line that was drawn by the capsid-binding compounds between two groups of rhinoviruses could be a reflection of genetic differences. In addition, who could assure us that the opposite spectra we were seeing for some compounds was not a curiosity, occurring for some exceptional serotypes?

It took more than a year and a great effort on the part of my people (Bart Dewindt and Jerry Snoeks) to do the tests—in triplicate—but finally we obtained an impressive set of IC_{50} data. We were now able to look at each individual serotype to see how susceptible they were to the different capsid-binding compounds. However, we wanted to extract the more intricate information that was present in these IC_{50} results. We found that the Spectral Map Analysis, an advanced multivariate analysis technique that had been developed by Dr. Paul Lewi of the JRF, was almost tailor-made for this purpose.

The result of Spectral Map Analysis is a two-dimensional plot which allows for a detailed graphic assessment of the observed differences in sensitivities of antiviral compounds (Figure 8). A virus is positioned in the plot based on its sensitivity for each of the 15 antivirals. When a virus has a more-than-average sensitivity to a given compound, it is attracted by that compound and vice versa. When a virus has a less-than-average sensitivity to a given compound, it is repelled from that compound. An antiviral is positioned in the same plot based on its

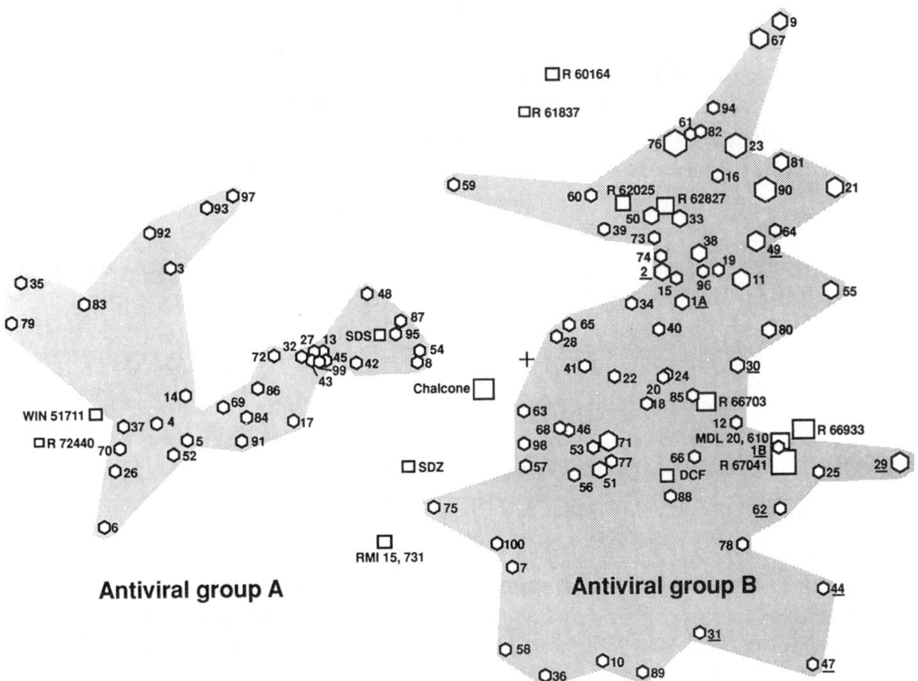

FIGURE 8. Spectral map obtained by multivariate analysis of antiviral tests. A panel of 15 capsid-binding compounds (for structures, see Andries et al., 1991) was tested against all rhinovirus serotypes. Hexagons represent serotypes; squares denote antiviral compounds. The positions of compounds are computed according to their specificities for the 100 rhinovirus serotypes and irrespective of their potencies. (Potency is defined here as the median reciprocal IC_{50} of a compound against the various viruses.) Viruses are located on the same map according to their specificities for the 15 compounds and irrespective of their sensitivities. (Sensitivity is defined here as the median reciprocal IC_{50} of a virus for the various compounds.) Median potencies of compounds and median sensitivities of serotypes are reflected by the sizes of hexagons and squares, respectively. The three-dimensional arrangement of compounds and serotypes has been rotated in order to show optimum separation of the two groups of serotypes. Reprinted with permission of Elsevier Science Publishers from Andries K et al. (1991): A comparative test of 15 compounds against all rhinoviruses as a basis for a more rational screening. *Antiviral Res* 16:213–225.

activity against each of the 100 rhinoviruses. All the interactions and positions are computed automatically, based on the IC_{50} data.

The analysis puts similar viruses, that is, viruses with similar susceptibilities to antiviral compounds, into clusters. We used cluster analysis to confirm the existence of two groups of rhinoviruses (Andries et al., 1990). The group on the left, which we called antiviral group A, consists of viruses having a more-than-average susceptibility to compounds such as WIN 51711. The group on the right, or antiviral group B, consists of viruses susceptible to antivirals such as R 61837, chalcone, and dichloroflavan.

The basis of the existence of the antiviral groups remains unknown. Given the fact that all these antivirals have a common binding site, our results indicate that this binding site must be somehow dimorphic. We know from the crystallographic studies of M. G. Rossmann's group that the WIN compounds bind into a long and narrow hydrophobic pocket in HRV14. The other rhinoviruses from antiviral group A, as well as the polioviruses, probably have a similar long and narrow pocket. On the other hand, the viruses from antiviral group B probably have a pocket that is different in composition and shape (perhaps shorter) to accept the generally shorter molecules that are active against these serotypes.

Relationship of Antiviral Groups to Amino Acid Sequence

We tried to find evidence for the dimorphic nature of the pocket or the amino acid lining the pocket by looking at the seven available sequences of rhinoviruses. We discovered the existence of a high correlation between the IC_{50} data assessed by Spectral Map Analysis, on the one hand (by measuring the distances between the points on the map), and the percentage of amino acid identities in the antivirals binding site, on the other. The closer two sequenced viruses lay to each other on the map, the more amino acids lining the pocket were found to be identical.

This indicated that the antiviral compounds had really worked as a panel of small molecular probes to characterize the composition of the antiviral binding sites of all these viruses, a process very similar to the characterization of epitopes by a panel of different monoclonal antibodies.

We tried to correlate our data set with other genome parts too. To our surprise, we also found high correlations with amino acid identities of VP1 and even of the whole P1 region. Our data suggest that the identities found in amino acid composition of the pocket are correlated with the identities in amino acid composition of the whole capsid and that mutations in the pocket are a reflection of random mutations in other parts of the genome. In other words, the Spectral Map data not only give one an idea as to the homology in pocket composition of unsequenced viruses but also appear to reflect the overall genetic diversity of all rhinoviruses.

Relationship of Antiviral Groups to Evolution of Rhinoviruses

Although our data, admittedly, are entirely obtained at the amino acid level and not at the nucleotide level, the model allows for a new hypothesis on the evolution of picornaviruses. It suggests that there was a common ancestor for group A rhinoviruses and polioviruses, because these viruses are computed to lie within the same antiviral group and therefore share a similar antiviral binding site. All antiviral group A viruses but one (HRV87) were shown to bind to the same cellular receptor, ICAM-1 (Uncapher et al., 1991; also see Chapter 9 in this volume). Next, a further divergence from group A rhinoviruses occurred, resulting in group B rhinoviruses with the same receptor specificity but a different antiviral binding site; and finally, a group of minor receptor–binding viruses branched off. If the divergence between the receptor groups had taken place before the divergence of the antiviral groups, minor receptor group viruses would have been distributed randomly over both antiviral groups, which is not the case. Viruses belonging to the minor receptor group are, without exception, all mapped in the same region of antiviral group B.

Relationship of Antiviral Groups to Occurrence of Clinical Colds

Viruses belonging to different antiviral groups even seemed to have a differential pathogenicity. We retrospectively gathered all available information on serotyped rhinovirus isolates of people with colds. During the last three decades, a total of 1205 isolates have been serotyped. Given the fact that antiviral group B contains about twice as many serotypes as antiviral group A, one would expect that serotypes of that group would account for about twice as many colds. However, this is not the case. Serotypes from group B accounted for five times as many colds as serotypes from group A, or 2.5 times as many per serotype. The chance that this was just coincidence was calculated to be less than 1 in 10,000. It looks as if rhinoviruses throughout evolution have become increasingly adapted to the nasal mucosa and have increased their ability to cause clinical disease. We therefore assume that something in the structure of group B viruses might endow them with a higher pathogenicity.

Use of Antiviral Groups to Rationalize Our Screening

It goes without saying that we also used the model to rationalize our search for better antivirals (Andries et al., 1991), although initially the rationale to do the analysis was more curiosity than anything else. From the analysis it can be deduced that screening of new antivirals against one or two serotypes can be very misleading, if that serotype is not representative for others. For instance, in the early days we used RV9 for screening. It is probably not a coincidence that we initially came up with a compound highly active against RV9 and some other serotypes located in the neighborhood of RV9 (Figure 8) but inactive against others. By using screening viruses that are located at the edges of the group, such as RV9, we had really diminished our chances of picking up interesting leads of broad-spectrum compounds. It is obviously necessary to select two viruses, preferentially one from the center of each antiviral group and both having a very high general sensitivity for antiviral compounds.

Once a lead compound has been found, it becomes increasingly important to learn as much as possible about its antirhinovirus spectrum. In order to do that, we selected a panel of 17 viruses from the Spectral Map. As selection criteria we took into account what was known about the serotypes (for instance, the sequence) and, more importantly, the position of that serotype in our model. Each of these 17 viruses can be seen as a representative for several other viruses. When a new variant of the pyridazinamine group is to be evaluated, it is tested against the whole panel of 17 serotypes. The calculation of a median IC_{50} from the IC_{50}'s obtained for the serotypes of each antiviral group allows for a very simple but accurate comparison of a compound's potency and antiviral spectrum (Table 1), especially as the values obtained with the panel of 17 turned out to be highly predictive for those obtained from the testing of all 100 serotypes (Andries et al., 1991). Thus it became possible for us to rationalize our screening system in search for broad-spectrum compounds.

Chemistry

One of the other approaches we followed to learn more about the existence of one or more binding sites was to synthesize what we called "hybrid" compounds, that is, compounds with substructures of WIN

TABLE 1. Antiviral activities of capsid-binding compounds[a]

HRV	AVG	R 61837	WIN 51711	R 72440	R 73884	Pirodavir
14	A	>32	0.17	0.7	0.005	0.011
42	A	>32	3.6	>32	7	>32
45	A	>32	1.5	>32	>32	1.6
70	A	>32	0.016	0.032	0.003	0.002
72	A	>32	1	4.05	0.016	0.143
86	A	>32	1	0.8	0.029	0.011
Median IC_{50}	A	>32	1	2.425	0.022	0.077
Median IC_{50}	B	1.6	2.3	>32	0.6	0.005
2	B	0.057	0.6	1.5	0.5	0.002
9	B	0.012	3.5	32	0.2	0.026
15	B	0.251	1.1	>32	0.9	0.031
29	B	0.070	1.6	12.8	0.4	0.003
39	B	1.0	2.4	>32	2.8	0.006
41	B	1.9	>32	>32	>32	0.291
51	B	6.8	2.3	0.331	0.005	0.005
59	B	5.7	32	>32	3.5	0.007
63	B	5.6	3.4	4	0.6	0.003
85	B	1.6	1.5	>32	0.6	0.002
89	B	2.7	1.6	>32	13	0.002
EC_{82}		>32	3.5	>32	7	0.143

[a] IC_{50}'s (in micrograms per milliliter) obtained with R 61837, R 72440, WIN 51711, R 73884, and pirodavir against 17 representative human rhinovirus (HRV) serotypes. Viruses are listed according to their antiviral group (AVG) allocation. The median IC_{50} for each antiviral group and the concentration effective in inhibiting 14 out of the 17 (82.3%) representative serotypes (EC_{82}) are also indicated.

51711 and pyridazinamines (Figure 9). This idea resulted from a brainstorm session with two chemists, Drs. Marcel Janssen and Raymond Stokbroekx, and our medicinal chemist, Dr. Henri Moereels. Our incentive was pure curiosity. What would the spectrum of such a hybrid compound look like? If there were two binding sites, we would not expect such a hybrid molecule to be active. If there was one binding site, we did not really know what to expect. To our amazement, one of the first three compounds that were made (R 72440, Figure 9) not only exerted antiviral activity, it also had a spectrum that was completely different from that of R 61837 and rather similar to that of WIN 51711 (Table 1). Although it was not a very potent compound, we knew we were on the right track.

Using the rational screening approach described above, we began SAR studies with about 500 chemical variations of our new "hybrid" lead compound, R 72440. Our intention at that time was the development of a compound highly active against antiviral group A serotypes. A combination of such a compound with one highly active against group B viruses would give us a broad-spectrum mixture. We did not dream of

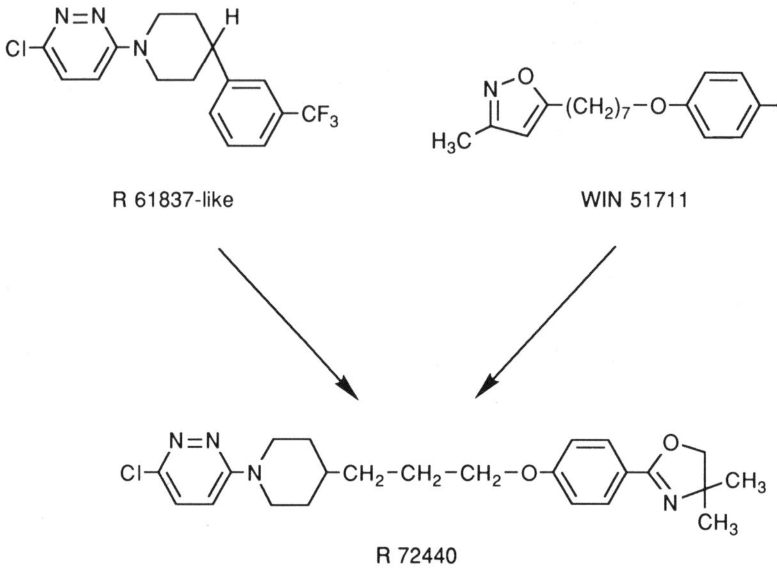

FIGURE 9. Synthesis of "hybrid" compound R 72440.

finding a real broad-spectrum compound, that is, a compound with activity against both groups of viruses. Ironically, that was exactly what we found. We could hardly believe our own test results when we saw the first compounds with a mixed spectrum (R 73884, Figure 10; see also Table 1). From then on, everything escalated, and our chemist eventually synthesized the compound coded R 77975, or pirodavir (Figure 10).

Inhibition of Picornaviruses by Pirodavir

Pirodavir, a substituted phenoxypyridazinamine, represents a new class of broad-spectrum antiviral drugs with significant in vitro efficacy against several members of the picornavirus family. Pirodavir is especially active against rhinoviruses, with MICs as low as 0.001 μg/mL (0.004 μM) for some serotypes. Compared with its predecessor R 61837, a greater than 500-fold improvement in potency was obtained, as shown by the drop in EC_{80} (calculated from the result of testing 100 rhinovirus serotypes) from >32 μg/mL for R 61837 to 0.064 μg/mL for R 77975 (Andries et al., 1992). The increase in potency was accompanied by a marked broadening of the spectrum. While R 61837 is almost exclusively active against rhinoviruses from antiviral group B, pirodavir is highly active against rhinoviruses from both antiviral groups (Table 1). Pirodavir but not R 61837 is also effective in inhibiting all enteroviruses

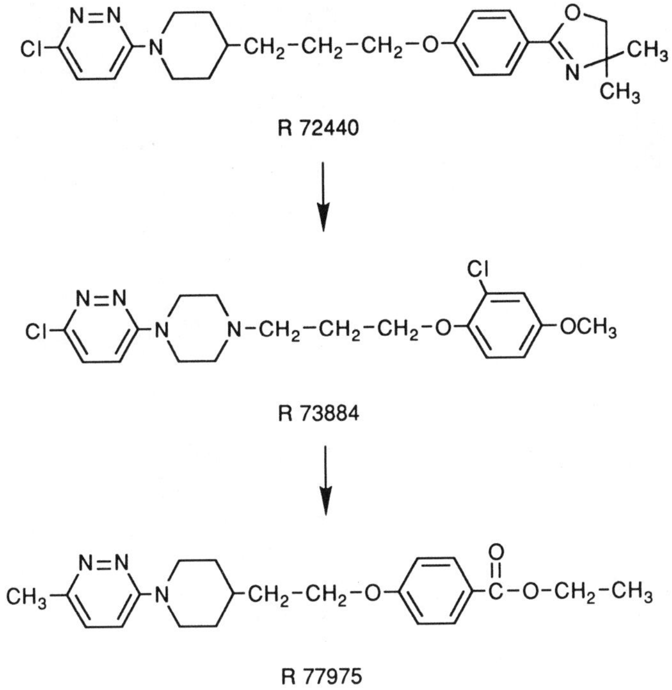

R 72440

R 73884

R 77975

FIGURE 10. Structures leading to R 77975 (pirodavir).

tested, although at higher concentrations (EC_{80} = 1.3) than those necessary to inhibit rhinoviruses (Figure 11). Close chemical analogues of pirodavir are better inhibitors of enteroviruses such as poliovirus, but at the same time they are less potent inhibitors of rhinoviruses.

The testing of a set of 17 representative serotypes provided predictive information for the results subsequently obtained after testing all rhinoviruses. A pirodavir concentration of 0.143 μg/mL, sufficient to inhibit 14 out of 17 (82.3%) screening serotypes (Table 1), also inhibited 84 out of 101 serotypes (83.1%).

The Binding Site of Pirodavir

We were still left with the question of how we could possibly explain the mere existence of a broad-spectrum compound after finding out that capsid-binding compounds are either active against antiviral group A or against group B, but never against both. The differential activity of most capsid-binding compounds toward serotypes from antiviral group A or B can be explained by their putative binding to particular amino acids

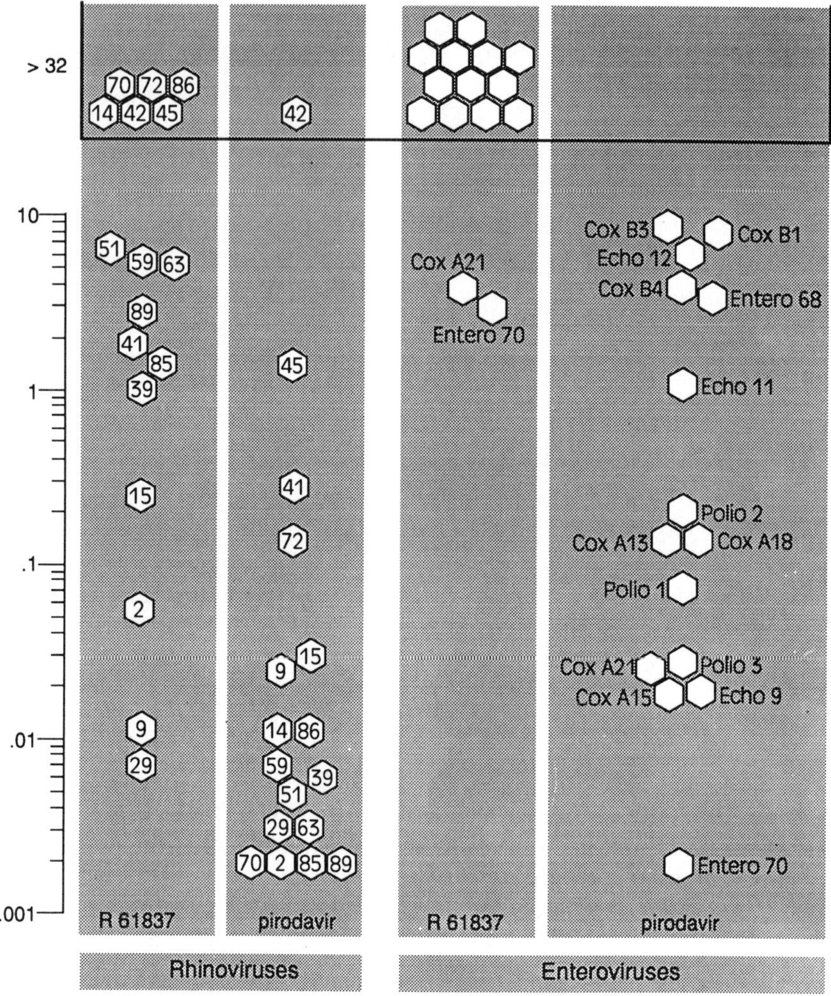

FIGURE 11. Antipicornavirus spectra of R 61837 and pirodavir. The IC_{50}'s (in micrograms per milliliter) of 17 representative rhinoviruses (labeled by their serotype number) and 15 enteroviruses (labeled if susceptible) were assayed. The IC_{50} is defined as the concentration required to reduce the inhibition of the formation of formazan from MTT by 50% (Andries et al., 1992).

lining the antiviral pocket, which are different in rhinoviruses belonging to a different antiviral group. Pirodavir, despite having a markedly different antiviral profile, apparently binds to the same antiviral target. Indeed, some HRV9 mutants that were resistant to R 61837 were found to be cross-resistant to pirodavir. Furthermore, preliminary results of X-ray crystallographic studies indicate that pirodavir also binds into the hydrophobic pocket of HRV14 (Dr. Michael G. Rossmann, personal communication).

The different antiviral profile—together with the sharing of the antiviral binding site—suggests that the broad-spectrum activity of the compound is achieved by its binding to amino acids from the antiviral pocket that are conserved throughout rhinoviruses belonging to both antiviral groups and enteroviruses. Apart from an asparagine at the back of the pocket, only two amino acids are conserved in every susceptible serotype sequenced thus far: two tyrosines at positions 152 and 197 in VP1 (Figure 7, HRV14 numbering). We strongly suspect that the broad spectrum activity of pirodavir can be explained by its binding to these conserved amino acid residues.

Resistance to Pirodavir

Although the potency and spectrum are impressive, some serotypes are naturally resistant to the compound. It is interesting to take a closer look at these. Of the initial panel of 17 viruses, two (HRV42 and HRV45) were resistant, that is, not susceptible to concentrations lower than 1 μg/mL (Table 1). Both serotypes appear at the same region (the right edge of antiviral group A) on the Spectral Map of serotypes and antiviral compounds (Figure 8).

Upon testing the other serotypes, we were amazed to see that all serotypes located in the neighborhood of these two serotypes were resistant and that only one other resistant virus could be found that was not positioned in that region. This finding considerably boosted our confidence in the system. It showed that from the results obtained with the 17 serotypes, a prediction can be made as to which serotypes are most likely to be resistant to a given compound.

The binding of narrow spectrum compounds such as R 61837 to nonconserved amino acids provides a rationale for why resistant mutants against such compounds can easily arise. In fact, in original virus stocks not exposed to any antiviral, resistant variants can easily be found. When such a population comes into contact with an antiviral, the susceptible variants are suppressed, so that a population with a higher percentage of resistant variants is selected for. There is no good reason to believe that these resistant variants would be less pathogenic, since resistant variants naturally occur in the form of other serotypes.

The probable binding of pirodavir to conserved amino acid residues can not only provide an explanation of the expanded spectrum but also implies the smaller possibility that resistant mutants will emerge. Indeed, amino acids can only be conserved when their presence facilitates the survival of the virus and their bsence renders the virus less viable.

When we tried to raise resistant mutants against pirodavir in vitro, we found some evidence in support of this hypothesis. Resistant variants of HRV Hank's (an untyped challenge strain) could be isolated, but only if the virus was replicated over a period of 1 week in the presence of a drug concentration within a very narrow "window": higher than the IC_{50} but at the same time lower than 20 times the IC_{50}. The level of resistance was always very low. For 22 out of 24 isolates studied the IC_{50} increased by a factor of 5 or less, and for the other two isolates it increased by a factor between 5 and 25. Less sensitive isolates displayed altered biological characteristics, in that they were clearly more sensitive to acid or heat inactivation, grew to lower titers in single-step and multiple-step replication cycles, and produced smaller plaques. The impaired growth properties suggest that resistant isolates may have a reduced virulence. Our findings therefore suggest an inverse relationship between the acquisition of broad-spectrum activity and the possibility of emergence of resistant mutants. Studies in other serotypes are needed to confirm this hypothesis.

Submission of an IND and First Clinical Trials with Pirodavir

In the autumn of 1988, it was decided to plan clinical trials with pirodavir. As further collaboration with the Common Cold Unit in England had regrettably become impossible because of the closure of the unit on the occasion of Dr. David Tyrrell's retirement, an IND file was required. Dr. Fred Hayden at the University of Virginia became our principal clinical investigator.

The toxicity of pirodavir is minimal. Repeated dosing for 3 months to rats up to 160 mg/kg was well tolerated. While the dose of 10 mg/kg was clean, dosing at 40 and 160 mg/kg resulted in slight renal toxicity. The histological changes of the urinary bladder and, associated herewith, the altered serum and urine variables were considered to be related to the hydroxypropyl-β-cyclodextrin vehicle. Repeated oral dosing with piro-davir in dogs for 3 months was well tolerated. Other than temporarily soft feces, no other adverse effects were observed at 2.5 mg/kg. Dosing at 10 and 40 mg/kg resulted in slight toxicity, but with no specific target organ for overdosing. Mutagenicity tests such as the Ames test and the micronucleus test in mice and tolerance studies such as the primary eye irritation test and primary dermal irritation study were all negative.

The pharmacokinetics of pirodavir in animals and humans are characterized by the rapid hydrolysis of the ethyl ester to the corresponding carboxylic acid metabolite after intravenous, oral, and intranasal administration. The metabolite of pirodavir exerts no antiviral activity. The plasma levels and kinetics of the metabolite in humans are similar to those in the animal species. In both rats and humans, steady-state plasma levels of the metabolite were reached after the third of six intranasal applications of pirodavir per day and were less than a quarter of those after a single oral dose.

The IND for pirodavir was filed in the summer of 1989, and Dr. F. Hayden managed to start the first clinical trials at the end of the same year. Although it is premature to draw final conclusions about the clinical efficacy of the compound, those first trials in volunteers, experimentally infected with HRV Hank's strain, basically confirmed the findings obtained with R 61837 (Hayden et al., 1992). Pirodavir is thus efficacious in the prevention of colds caused by rhinoviruses. A therapeutic trial failed to demonstrate clinical efficacy, although the monitoring of virus excretion in nasal excretions demonstrated an effect on viral replication. The infectiousness of nasal secretions of people treated with pirodavir is significantly reduced as a result of the effect on viral replication and as the result of an additional virus neutralization effect of pirodavir. The reduced infectiousness of the secretions may possibly lead to a smaller chance of virus transmission, but this is subject for further study.

Conclusion and Future Prospects

Where do we stand now in the development of an effective drug against the common cold? Significant progress has been made in tackling each of the five obstacles mentioned in the introduction. We, as well as others (Woods et al., 1989), have come a long way in the development of broad-spectrum antipicornavirus agents. Clinical efficacy in humans has been achieved through the use of a new vehicle in a nasal formulation. There are also indications that the problem of emergence of resistant mutants may be less important than previously anticipated. However, despite some real scientific breakthroughs, the ultimate goal—the achievement of therapeutic clinical efficacy—has not yet been met.

Is further progress toward this goal possible? One can hardly expect that a further substantial increase in the potency of pirodavir is within easy reach. After all, an increase in the strength of the hydrophobic bonds between capsid-binding compounds and hydrophobic amino acids in the target site is bound to be limited by the available contact surface area. Furthermore, once a high broad-spectrum potency is achieved, a further increase in potency for one virus type may well be accompanied by a decrease in potency for others. The use of combinations of compounds with synergistic effects might be considered. However, the fact that synergistic effects appear to be different from serotype to serotype (Al-Nakib and Tyrrell, 1987) implies that the development of a broad-spectrum synergistic mixture will be far from easy.

The other possible approach to increasing the clinical efficacy of capsid-binding compounds is the further improvement of an intranasal formulation. The use of a vehicle that slowly releases a solution of the antiviral compound in the nasal environment may significantly increase the antiviral activity of an existing compound.

Although capsid-binding compounds might not yet have fulfilled their promises, their development has taught us a lot about the picornaviruses themselves. They turn out to be very useful tools for studying the early interactions of these viruses with their host cells in great detail. A further slightly unexpected spin-off of the research on capsid-binding compounds is their possible use as vaccine stabilizers. Indeed, many of the compounds described above that inhibit the replication of polioviruses at the same time stabilize the poliovirus coat proteins and are thus potentially useful as stabilizers of inactivated poliovirus vaccines (Rombaut et al., 1991).

Acknowledgments

Although many collaborators and colleagues made a significant contribution to the work presented herein, and I am very grateful to them, I would like to dedicate this chapter to Dr. Paul Janssen, the driving force behind every important event in this and many other projects. There are no bounds to his enthusiasm and his personal encouragement of our scientific curiosity.

References

Al-Nakib W, Tyrrell DAJ (1987): A "new" generation of more potent synthetic antirhinovirus compounds: Comparison of their MICs and their synergistic interactions. *Antiviral Res* 8:179–882

Al-Nakib W, Higgins PG, Barrow GI, Tyrrell DAJ, Andries K, Vanden Bussche G, Taylor N, Janssen PAJ (1989): Suppression of colds in human volunteers challenged with rhinovirus by a new synthetic drug (R61837). *Antimicrob Agents Chemother* 33:522–525

Andries K, Dewindt B, De Brabander M, Stokbroekx R, Janssen P (1988): *In vitro* activity of R 61837, a new antirhinovirus compound. *Arch Virol* 101:155–67

Andries K, Dewindt B, Snoeks J, Willebrords R (1989): Lack of quantitative correlation between inhibition of replication of rhinoviruses by an antiviral drug and their stabilization. *Arch Virol* 106:51–61

Andries K, Dewindt B, Snoeks J, Wouters L, Moereels, H, Lewi PJ, Janssen PAJ (1990): Two groups of rhinoviruses revealed by a panel of antiviral compounds present sequence divergence and differential pathogenicity. *J Virol* 64:1117–1123

Andries K, Dewindt B, Snoeks J, Willebrords R, Stokbroekx R, Lewi, PJ (1991): A comparative test of 15 compounds against all rhinoviruses as a basis for a more rational screening. *Antiviral Res* 16:213–225

Andries K, Dewindt B, Snoeks J, Willebrords R, Van Eemeren K, Stokbroekx R, Janssen PAJ (1992): *In vitro* activity of pirodavir (R 77975), a substituted phenoxy-pyridazinamine with broad-spectrum antipicornaviral activity. *Antimicrob Agents Chemother* 36:100–107

Anonymous (1984): Enviroxime. *Drugs Future* 9:221–222

Barrow GI, Higgins PG, Tyrrell DAJ, Andries K (1990): An appraisal of the efficacy of the antiviral R 61837 in rhinovirus infections in human volunteers. *Antiviral Chem Chemother* 1:279–285

Bauer DJ, Selway JWT, Batchelor JF, Tisdale M, Caldwell IC, Young DAB (1981): 4′,6-Dichloroflavan (BW683C), a new anti-rhinovirus compound. *Nature (London)* 292:369–370

Hayden FG, Andries K, Janssen PAJ (1992). Safety and efficacy of intranasal pirodavir (R 77975) in experimental rhinovirus infection. *Antimicrob Agents Chemother* 36:727–732

Ishitsuka H, Ninomiya YN, Ohsawa C, Fujiu M, Suhura Y (1982): Direct and specific inactivation of rhinovirus by chalcone Ro 09–410. *Antimicrob Agents Chemother* 22:617–621

Otto MJ, Fox MP, Fancher MJ, Kuhrt MF, Diana GD, McKinlay MA (1985): *In vitro* activity of WIN 51711, a new broad-spectrum antipicornavirus drug. *Antimicrob Agents Chemother* 27:883–886

Rombaut B, Andries K, Boeyé A (1991): A comparison of WIN 51711 and R 78206 as stabilizers of poliovirus virions and procapsids. *J Gen Virol* 72:2153–2157

Rossmann MG, Arnold E, Erickson JW, Frankenberger EA, Griffith JP, Hecht H, Johnson JE, Kamer G, Luo M, Mosser AGT, Rueckert RR, Sherry B, Vriend G (1985): Structure of a human common cold virus and functional relationship to other picornaviruses. *Nature (London)* 317:145–153

Rueckert RR (1990): Picornaviridae and their replication. In *Virology*, Fields BN, Knipe DM, eds. New York: Raven Press

Smith TJ, Kremer MJ, Lou M, Vriend G, Arnold E, Kamer G, Rossmann MG, McKinlay MA, Diana GD, Otto MJ (1986): The site of attachment in human rhinovirus 14 for antiviral agents that inhibit uncoating. *Science* 233:1286–1293

Uncapher CR, DeWitt CM, Colonno RJ (1991): The major and minor receptor group families contain all but one human rhinovirus serotype. *Virology* 180:814–817

Woods MG, Diana GD, Rogge MC, Otto MJ, Dutko FJ, McKinlay MA (1989): In vitro and in vivo activities of WIN 54954, a new broad-spectrum antipicornavirus drug. *Antimicrob Agents Chemother* 33:2069–2074

9

sICAM-1 as a Receptor Antagonist for Rhinoviruses: A Model System of Adhesion Molecules as Cell Receptors for Viruses

STEVEN D. MARLIN

Virus Receptors as Molecular Targets

As obligate intracellular parasites, viruses must subvert normal cellular machinery in order to replicate themselves. However, the cell membrane represents a formidable barrier that prevents viral access to that intracellular machinery. In some cases, viruses can penetrate that barrier via nonspecific means, for example, by passive engulfment during pinocytosis. More commonly, viruses begin entry into cells by actively binding to cell surface structures that allow virus attachment and subsequent penetration into the cell. Potentially, any component of the cell membrane can serve as a virus receptor, including proteins, carbohydrates, and lipids. Either a single structure, or multiple molecules may be used as receptors by the same virus.

As targets for therapeutic intervention of viral diseases, virus receptors offer some distinct advantages and disadvantages. Since receptor binding is the first event in infection, blocking binding actually prevents infection. If complete blocking can be achieved, preventing infection is clearly more desirable than attempting to disrupt an ongoing process. A receptor antagonist does not have to, and in fact should not, enter the cytoplasm of the cell. Thus, drug delivery could be greatly simplified. Moreover, an extracellular antagonist may be less likely to interfere with intracellular metabolism and create toxic side effects. The generation of virus mutants resistant to drug therapy, or variation between virus strains, is often a limitation or concern in antiviral therapy. At face value, virus receptor binding sites seem to be highly conserved and thus less likely to mutate to a drug-resistant phenotype. For viruses that utilize a single, obligate receptor, it has been hypothesized that mutations in the

The Search for Antiviral Drugs
Julian Adams and Vincent J. Merluzzi, Editors
© 1993 Birkhäuser Boston

receptor binding site would be lethal to the virus and therefore rare or nonexistent.

There are potential limitations to receptor-targeted therapy as well. Even complete inhibition of a specific virus–receptor interaction may be insufficient to prevent infection if the virus is capable of utilizing alternate receptors or can productively enter the cell through nonspecific means. Unlike viral enzyme targets where a limited copy number of catalytic molecules must be inhibited, virus receptors probably exist in large numbers on the cell. Thus, receptor-targeted therapy may require high amounts of drug to saturate all available targets. Finally, a receptor-targeted therapeutic agent runs the risk of adverse physiological consequences if it inhibits the normal cellular function of the receptor.

Cell Adhesion Molecules as Virus Receptors

The list of proteins identified as virus receptors is surprisingly short.

Rhinovirus	*ICAM-1*
HIV, SIV	*CD4*
Epstein–Barr Virus	*CR2 (C3dR)*
CMV, Adenovirus, Semliki	*HLA Class I*
LDH Virus	*HLA Class II*
Vaccinia	*EGF Receptor*
Reovirus	*Adrenergic Receptor*
Rabies Virus	*Acetycholine Receptor*
Hepatitis B Virus	*Poly IgA Receptor*
Rad. Leukemia Virus	*TCR/T4 Complex*
Foot-and-Mouth Disease	*Integrin ??*
Poliovirus	*Ig Supergene Family*
Polyomavirus	
Influenza	
Sendai	*Sialyloligosaccharides*
Newcastle Disease Virus	
VSV, Rabies	*Lipids*

Of those proteins, perhaps two bona fide "cell adhesion molecules" (CAMs) have been proven to be virus receptors: CD4 for HIV, and ICAM-1 (intercellular adhesion molecule-1) for certain rhinoviruses. However, the poliovirus receptor is a member of the immunoglobulin supergene family and contains structural motifs commonly used by CAMs. In addition, the receptor for foot-and-mouth disease virus has been implied to be an integrin based on inhibition with peptides containing the RGD (arginine–glycine–aspartic acid) sequence found in many integrin receptor ligands. Thus, perhaps four CAMs [more if human leukocyte antigens (HLA) class I and class II are considered to be CAMs] are used as virus receptors. This represents a substantial proportion of the total number of cell surface proteins identified as virus receptors.

The question then arises whether CAMs are particularly "attractive" for subversion by viruses compared to other cell surface proteins. Some of the characteristics of CAMs would in fact seem to make them ideal for use as virus receptors. First, as mediators of cell interaction with either other cells or the external environment (e.g., the extracellular matrix), they are grossly accessible on the cell surface, often present in large copy number, and may represent a large percentage of the protein molecules present on the cell membrane. Secondly, the molecules themselves are structurally tailor-made for mediating macromolecular interactions with a significant binding affinity. The hypothesis that CAMs are particular targets for use as receptors by pathogens is bolstered by the observation that ICAM-1 is not only a rhinovirus receptor but also a receptor used in the attachment of malaria-infected red blood cells to endothelial cells (Berendt et al., 1989; Ockenhouse et al., 1992).

However, other aspects of adhesion molecules may limit their use as virus receptors. Many CAMs have a specific and limited cell/tissue distribution, whereas others are expressed only transiently. Thus, these adhesion molecules would present limited tissue targets, or restricted temporal opportunities for a virus to infect.

Teleological arguments aside, it remains to be seen whether CAMs are utilized as virus receptors more frequently than other kinds of cell surface molecules. If nothing else, the sheer number of CAMs and the intense research devoted to them almost guarantee that more CAMs will be identified as virus receptors.

ICAM-1 as a Model System

ICAM-1 is a CAM with a central role in mediating cell–cell adhesion in immune and inflammatory responses (reviewed in Springer, 1990; Hogg, 1991). Recently, ICAM-1 has been proven to be the receptor for a group of picornaviruses that can be causative agents of the common cold (Staunton et al., 1989; Greve et al., 1989). Although several different groups of viruses are responsible for the symptoms loosely described as a cold, approximately 50% are attributed to rhinoviruses. About 90% of the rhinoviruses have been classified as "major group" viruses since they utilize the same receptor (now known to be ICAM-1). Thus, perhaps 40% of colds are caused by major group rhinoviruses using ICAM-1 as a receptor. For these viruses, ICAM-1 seems to be the only means of entering and infecting cells (Figure 1). The evidence that ICAM-1 is, in fact, the receptor for these viruses is derived from gene transfection, antibody inhibition, and binding to purified protein, as well as comparisons of the independent characteristics of ICAM-1 and

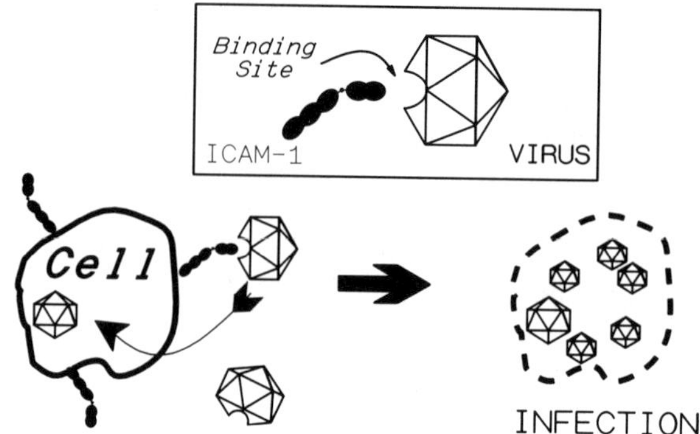

FIGURE 1. ICAM-1 is the receptor for the major group of rhinoviruses.

the "rhinovirus receptor" (Tomassini and Colonno, 1986; Staunton et al., 1989; Greve et al., 1989; Marlin et al., 1990).

The structure of ICAM-1 is consistent with its use as a rhinovirus receptor. Based on sequence analysis, molecular modeling, and analogies with homologous proteins, ICAM-1 is a membrane glycoprotein composed of five extracellular immunoglobulin (Ig) domains, a typical transmembrane domain, and a short cytoplasmic domain (Figure 2).

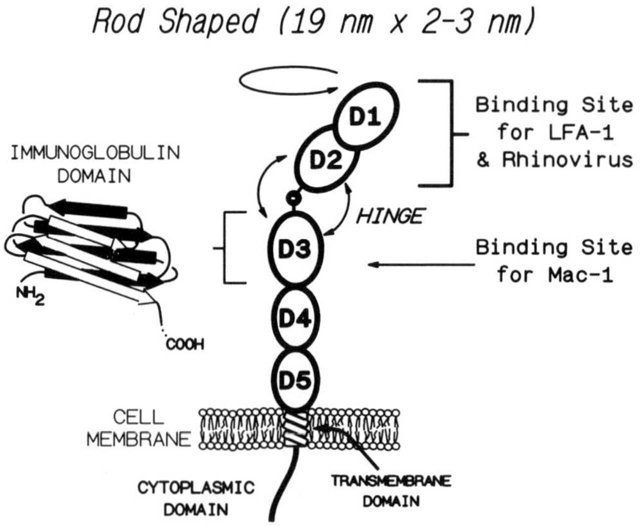

FIGURE 2. Idealized structure of ICAM-1. The protein is a transmembrane glycoprotein with five extracellular immunoglobulin domains connected by a flexible hinge. The structure is derived from sequence analysis, homology to other proteins, molecular modeling, and electron microscopy.

Electron micrographs of the purified protein indicate that it is a rod-shaped molecule with the Ig domains stacked one on top of another. The overall dimensions are approximately 19 nm by 2–3 nm. Both electron microscopy and sequence analysis suggest that the molecule has a flexible bend, or hinge, probably between domains D2 and D3.

By recombinant techniques, a soluble form of the normally membrane-bound ICAM-1 has been constructed and purified to homogeneity (Figures 3 and 4) (Marlin et al., 1990). Soluble ICAM-1 (sICAM-1) was constructed by introducing an in-frame stop codon just prior to the predicted transmembrane domain in the ICAM-1 gene. When the resulting protein is expressed in transfected cells, it lacks the transmembrane and cytoplasmic domains and as a result is secreted into the culture medium. The secreted protein is readily soluble in aqueous solution and can be easily purified.

sICAM-1 is a potent and specific inhibitor of rhinovirus infection, with half-maximal inhibition in cytopathic effect (CPE) assays of approximately 10 μg/mL (200 nM) (Figure 5). The inhibition is specific for viruses of the major group rhinovirus family and has no effect on minor group rhinoviruses or other, unrelated DNA or RNA viruses (Figure 6). As expected, the mechanism of action appears to be competitive inhibition of the binding of virus particles to cells: sICAM-1 is able to competitively inhibit the binding of radiolabeled virus particles (Figure 7) and under normal circumstances is readily reversible (data not shown).

The ability of sICAM-1 to inhibit rhinovirus infection indicates that sICAM-1 itself, or a derivative, could be a potential therapeutic agent. As a large complex protein of 452 amino acids), sICAM-1 must be produced via eukaryotic—and perhaps prokaryotic—expression systems. As a derivative of a normal human protein, sICAM-1 would be

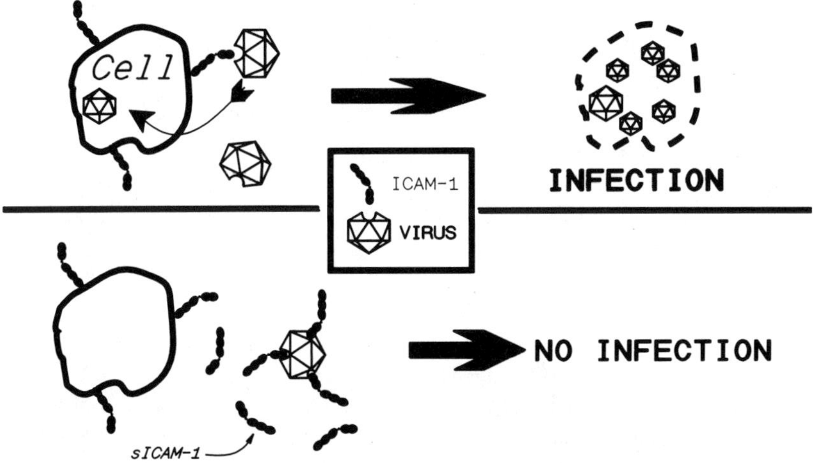

FIGURE 3. A soluble form of ICAM-1 inhibits rhinovirus infection.

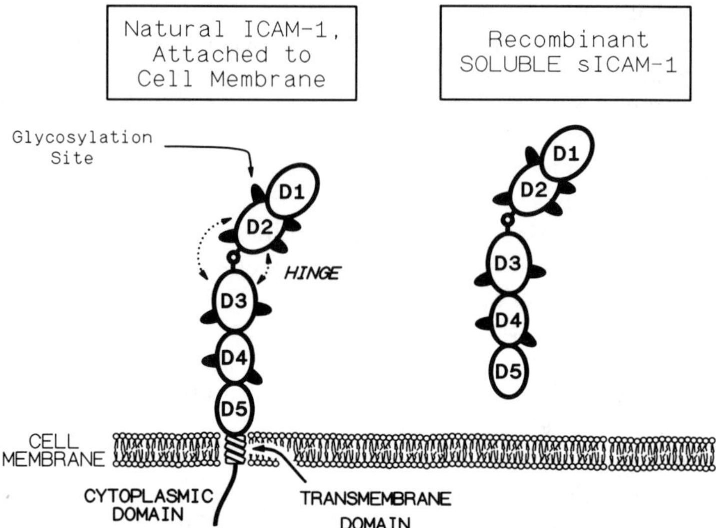

FIGURE 4. Idealized structure of membrane ICAM-1 and sICAM-1.

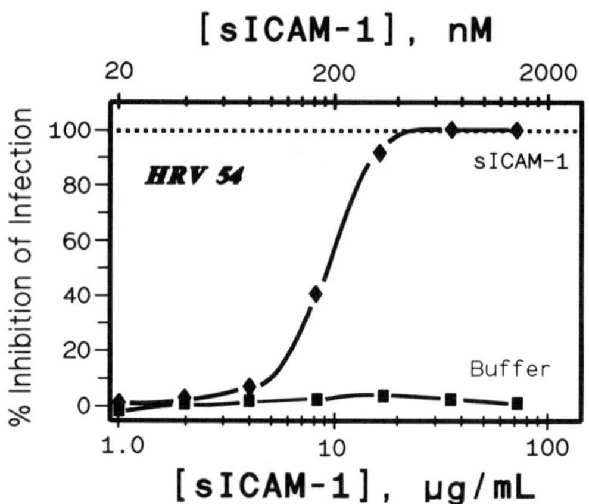

FIGURE 5. sICAM-1 inhibits rhinovirus infection. The ability of sICAM-1 to inhibit infection of HeLa cells by rhinovirus strain 54 was measured by CPE assay. An equivalent dilution of buffer from the same chromatographic purification was used as a control.

predicted to have little, if any, inherent toxicity; in fact, a form of sICAM-1 termed cICAM-1 can be detected circulating in the serum of healthy individuals (Rothlein et al., 1991). This expected lack of toxicity is an appealing characteristic for an agent targeted for treatment of a benign, recurring disease such as the common cold. However, a much

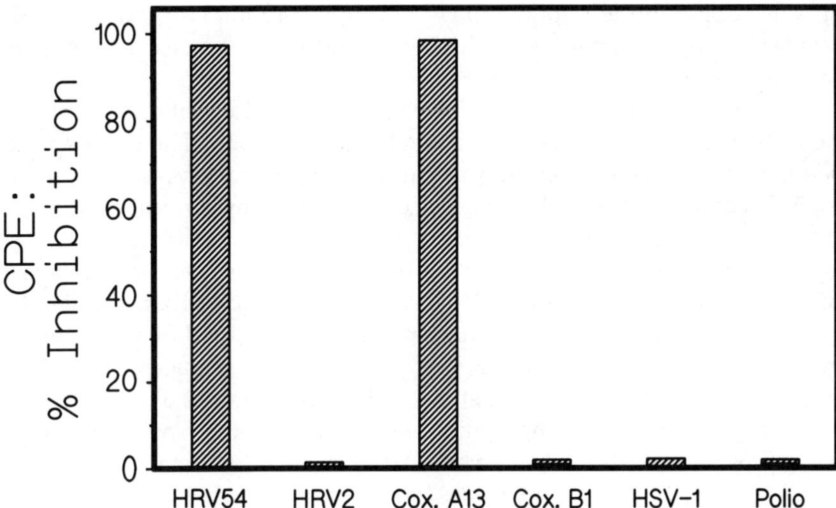

FIGURE 6. Virus specificity of sICAM-1. Purified sICAM-1 (5 µg/mL) was tested in CPE assays against the following viruses: HRV54, major group rhinovirus; HRV2, minor group rhinovirus; Cox. A13, coxsackie A13, a picornavirus that uses the major group receptor; Cox. B1, coxsackie B1, which does not use the major group receptor; HSV-1, herpes simplex virus 1; Polio, poliovirus I.

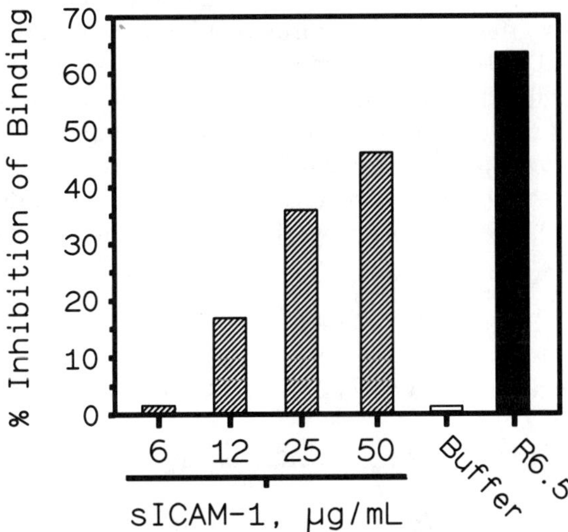

FIGURE 7. sICAM-1 inhibits rhinovirus binding. The binding of radiolabeled human rhinovirus particles (HRV14) to HeLa cells was measured in the presence of sICAM-1, a chromatography-buffer control, or the anti-ICAM-1 monoclonal antibody R6.5.

smaller polypeptide/peptide or a synthetic organic compound with similar binding properties could have production or cost advantages over the large protein.

In principle, the binding site for rhinovirus could reside in the polypeptide, the carbohydrate, or in a combination of both. The polypeptide sequences involved in binding to rhinovirus and to LFA-1* (a physiological receptor for ICAM-1) have been identified by a number of techniques, including antibody binding analysis, site-specific point mutagenesis, truncation of the molecule, domain deletions, and the analysis of molecular chimeras (Figure 8). Monoclonal antibodies that inhibit both rhinovirus and LFA-1 have been identified and mapped by point mutagenesis to domains D1 and D2, and thus strongly implicate those two domains in binding (Staunton et al., 1990; Lineberger et al., 1990). Chimeric molecules containing combinations of domains from the mouse and human molecules have also been used to identify domains critical for binding, since human rhinoviruses bind to human ICAM-1 but not to the structurally related murine homologue. As depicted in Figure 9, only chimeras containing the human versions of domain D1 (and to some extent D2) retained the ability to bind rhinovirus (McClelland et al., 1991; Staunton et al., 1990). Similarly, only molecules containing human D1 and D2 could bind to LFA-1 (Staunton et al., 1990). Thus, the binding sites for both rhinovirus and LFA-1 appear be located in domain D1. Although these sites overlap, they are not identical. As depicted in Figure 10, an extensive series of site-specific mutations in domains D1 and D2 has shown that mutations in some amino acids affect both binding sites, whereas others affect only rhinovirus or LFA-1 (Staunton et al., 1990; McClelland et al., 1991). In addition, point mutations in regions predicted to be topologically distinct can destroy binding, demonstrating that the site is complex and nonlinear. Although the binding site for rhinovirus is located in domain D1, that single domain is not sufficient; when truncated, secreted molecules are created, the D1 molecule does not bind, while the D1–D2 molecule is active (Figure 11) (Lineberger et al., 1990).

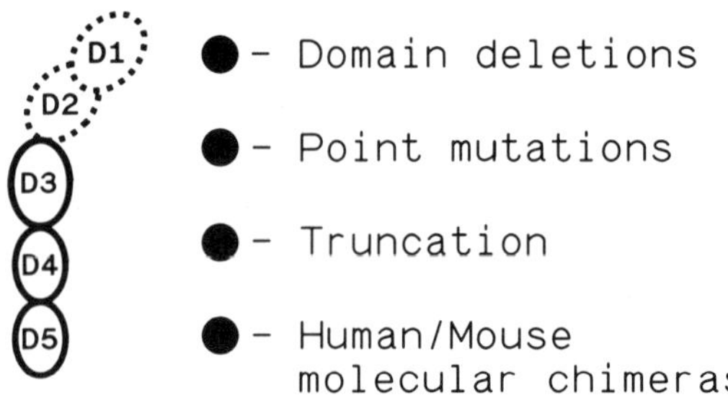

FIGURE 8. The binding sites for RV and LFA-1 are located in the first two domains. From Staunton et al. (1990), McClelland et al. (1991), and Register et al. (1991).

---

*LFA-1 = lymphocyte function-associated antigen-1.

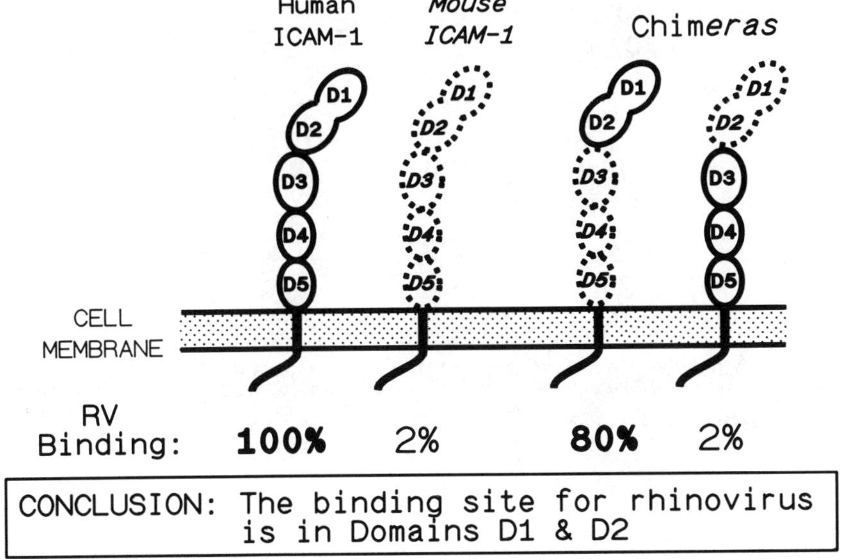

FIGURE 9. Activity of mouse/human chimeras. From Staunton et al. (1990).

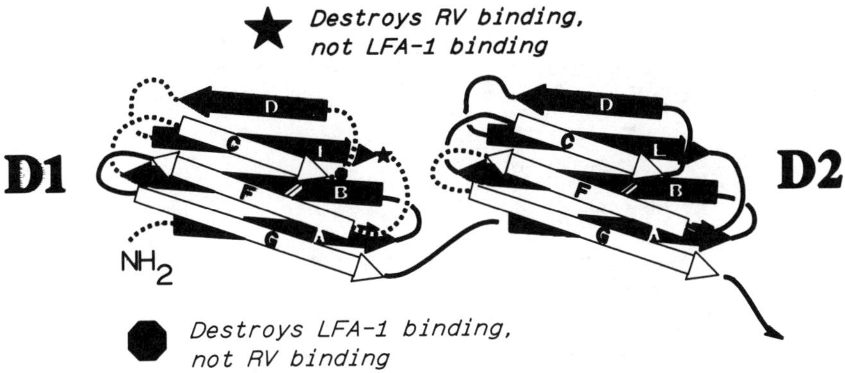

FIGURE 10. Mutations in ICAM-1 that affect RV binding. Stippled loops localize mutations that reduce or destroy such RV binding. From Staunton et al. (1990).

Thus, although both the rhinovirus and LFA-1 binding sites overlap, they are clearly distinct at the molecular level. This distinction raises the possibility that sICAM-1 or a mimic could be engineered to specifically bind to rhinovirus without binding to normal physiological receptors.

The location of the rhinovirus and LFA-1 binding site(s) predominantly in domain D1 is conceptually appealing. The extended rod shape of the molecule would orient domain D1 well beyond the net negative charge of the cell surface and thus provide increased accessibility for

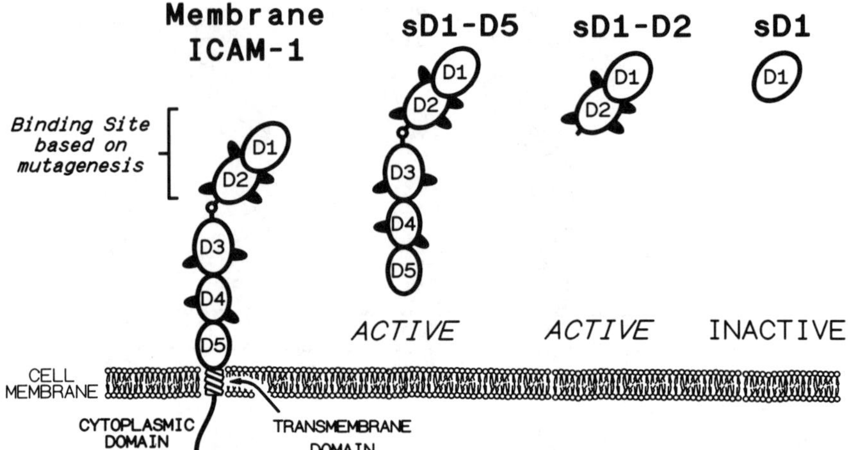

FIGURE 11. The first two domains are sufficient for rhinovirus inhibition.

binding. Moreover, the putative molecular flexibility contributed by the hinge after domain D2 could increase binding kinetics by increasing the apparent rate of diffusion of the binding site in space. The localization of the rhinovirus binding site to domain D1 is also consistent with the hypothesized binding site on the rhinovirus capsid. Based on high-resolution X-ray crystallography and virus mutation analysis, the binding site is a narrow "canyon"-like structure (Figures 12 and 13)

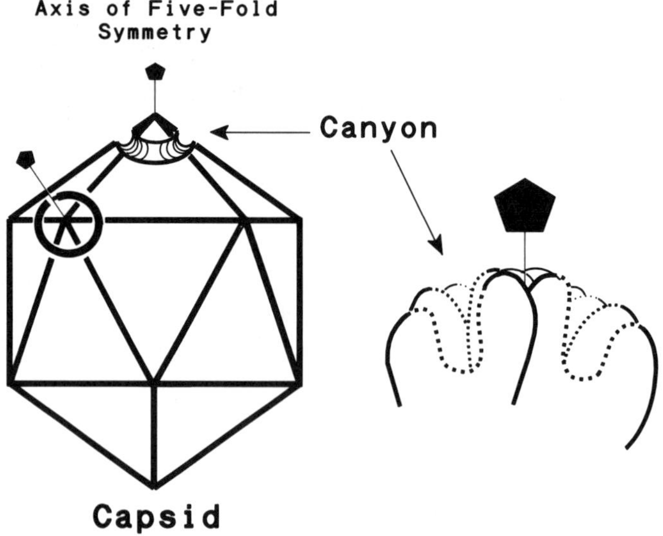

FIGURE 12. Rhinovirus capsid structure depicting canyon feature at each axis of fivefold symmetry. Adapted from Rossmann (1989): *Viral Immunol* 2:143–161.

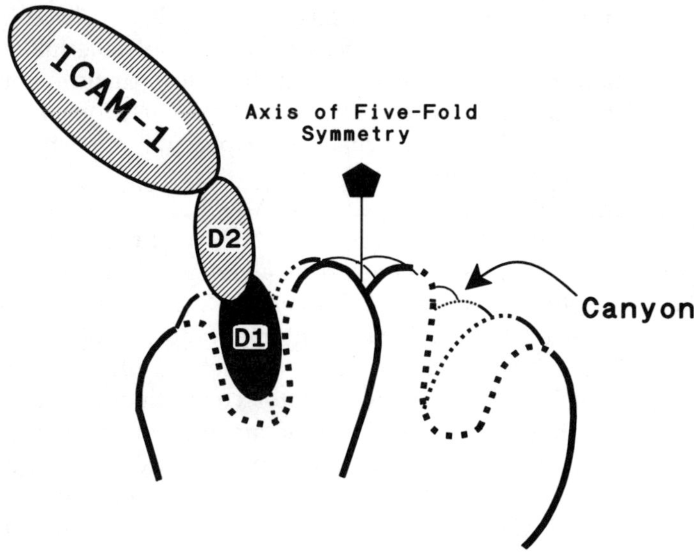

FIGURE 13. sICAM-1 binding to rhinovirus canyon. Adapted from Rossmann (1989): *Viral Immunol* 2:143–161.

(Rossmann, 1989; Rossmann and Palmenberg, 1988). The dimensions of this canyon preclude access to antibodies and may thus prevent an effective immune response against the virus-binding site. However, molecular modeling of ICAM-1 domain D1 indicates that this domain is of the correct dimensions to bind within the canyon (Giranda et al., 1990). Thus, both analysis of ICAM-1 and its site of interaction on the virus capsid are consistent with binding via domain D1. There appears to be some contribution from domain D2 as well, and the two domains may be interacting to form a single critical structure. Not only are these two domains involved in binding, but they are also sufficient without the rest of the ICAM-1 molecule: a soluble two-domain molecule (sD1–D2) is approximately as active as sICAM-1 itself in inhibiting rhinovirus (Figure 11) (Lineberger et al., 1990). A reduction of the molecule to an even smaller active moiety awaits further research.

As expressed in most cells, ICAM-1 is heavily glycosylated. Carbohydrate can account for as much as 40% of the total molecular mass, and there are eight potential N-linked glycosylation sites (notably, there are no glycosylation sites in domain D1 where the molecule is predicted to dock into the rhinovirus canyon). The contribution of the carbohydrate to binding then becomes an important question. Non-glycosylated sICAM-1 has been produced and purified from cells grown in the presence of tunicamycin, an inhibitor of N-linked glycosylation. Non-glycosylated sICAM-1 has equal activity in inhibiting rhinovirus infection as glycosylated sICAM-1. Thus, the carbohydrate is not critical to the activity of the molecule (although it could contribute other functions or contribute to structural stability). The lack of requirement for carbohydrate makes it probable that an active form of sICAM-1 can be made in expression systems such as insect cells, yeast, or bacteria.

As a potential therapeutic agent, sICAM-1 faces significant challenges. Considering the self-limiting, mild nature of the common cold, any agent used for this disease must have a very good safety profile and must be relatively inexpensive. Whether sICAM-1 will be useful for prophylactic use or for therapeutic intervention or both during an ongoing cold remains to be seen. Perhaps most significantly, since the mechanism of sICAM-1 is clearly receptor competition, it could only target perhaps those 40% of colds that are caused by major group rhinoviruses. Finally, as with any protein-based drug, creating a stable formulation and a method of delivery could be a hurdle. In particular, the normal rapid clearance mechanisms of the nose may require methods to maximize intranasal retention.

Apart from considerations as a therapeutic agent itself, sICAM-1 offers significant opportunities as a molecular tool for research and drug discovery. Perhaps most exciting is the prospect of resolving the three-dimensional structure of ICAM-1 and accurately defining the molecular interactions with the rhinovirus capsid (the structure of the virus is already known). Having the three-dimensional structure of both the virus and its receptor may present unique opportunities for de novo "rational" design of small molecule inhibitors with unique properties.

sICAM-1 also has utility as a probe in the search for small molecule inhibitors via the traditional approach of large-scale "random" screening. sICAM-1 can be labeled by biotinylation of the carbohydrates without apparently altering its inhibitory activity or specificity. Biotinylated sICAM-1 can thus be used in an ELISA-based, cell-free screen to search for new inhibitors. In a simple modification of the same technology, biotinylated sICAM-1 can be used to detect and identify major group rhinoviruses (Last-Barney et al., 1991). In principle, similar approaches could be used with any virus receptor to generate screening or diagnostic assays.

In summary, there appear to be valid reasons for expecting that cell adhesion molecules will often be used as receptors for viruses (or other pathogens). For antiviral therapy, targeting the virus receptor or the counterstructure on the virus offers some unique and new advantages in preventing virus infection. As a model, the rhinovirus/ICAM-1 system illustrates the opportunities for fruitful basic research combined with potential drug discovery.

References

Berendt AR, Simmons DL, Tansey J, Newbold CI, Marsh K (1989): Intercellular adhesion molecule-1 is an endothelial cell adhesion receptor for *Plasmodium falciparum*. *Nature (London)* 341:57–59

Giranda VL, Chapman MS, Rossmann MG (1990): Modeling of the human intercellular adhesion molecule-1, the human rhinovirus major group receptor. *Proteins* 7:227–233

Greve JM, Davis G, Meyer AM, Forte CP, Yost SC, Marlor CW, Kamarck ME, McClelland A (1989): The major human rhinovirus receptor is ICAM-1. *Cell (Cambridge, MA)* 56:839–847

Hogg N (1991): Integrins and ICAM-1 in immune responses. An integrin overview. *Chem Immunol* 50:1–12

Last-Barney K, Marlin SD, McNally EJ, Cahill C, Jeanfavre D, Faanes RB, Merluzzi VJ (1991): Detection of major group rhinoviruses by soluble intercellular adhesion molecule-1 (sICAM-1). *J Virol Methods* 35:255–264

Lineberger DW, Graham DJ, Tomassini JE, Colonno RJ (1990): Antibodies that block rhinovirus attachment map to domain 1 of the major group receptor. *J Virol* 64:2582–2587

Marlin SD, Staunton DE, Springer TA, Stratowa C, Sommergruber W, Merluzzi VJ (1990): A soluble form of intercellular adhesion molecule-1 inhibits rhinovirus infection. *Nature (London)* 344:70–72

McClelland A, DeBear J, Yost SC, Meyer AM, Marlor CW, Greve JM (1991): Identification of monoclonal antibody epitopes and critical residues for rhinovirus binding in domain 1 of intercellular adhesion molecule 1. *Proc Natl Acad Sci USA* 88:7993–7997

Ockenhouse CF, Gaur A, Diamond MS, Betageri R, Springer TA, Staunton DE (1992): *Plasmodium falciparum*-infected erythrocytes bind ICAM-1 as a site distinct from LFA-1, Mac-1, and human rhinovirus. *Cell (Cambridge MA)* 68:63–69

Register (1991): *J Virol* 65:6589

Rossmann MG (1989): The canyon hypothesis. *Viral Immunol* 2:143–161

Rossmann MG, Palmenberg AC (1988): Conservation of the putative receptor attachment site in picornaviruses. *Virology* 164:373–382

Rothlein R, Mainolfi EA, Czajkowski M, Marlin SD (1991): A form of circulating ICAM-1 in human serum. *J Immunol* 147:3788–3793

Springer TA (1990): Adhesion receptors of the immune system. *Nature (London)* 346:425–434

Staunton DE, Merluzzi VJ, Rothlein R, Barton R, Marlin SD, Springer TA (1989): A cell adhesion molecule, ICAM-1, is the major surface receptor for rhinoviruses. *Cell (Cambridge MA)* 56:849–853

Staunton DE, Dustin ML, Erickson HP, Springer TA (1990): The arrangement of the immunoglobulin-like domains of ICAM-1 and the binding sites for LFA-1 and rhinovirus. *Cell (Cambridge MA)* 61:243–254

Tomassini JE, Colonno RJ (1986): Isolation of a receptor protein involved in attachment of human rhinoviruses. *J Virol* 58:290–295

10

Inactivation of Herpes Simplex Virus Ribonucleotide Reductase by Subunit Association Inhibitors: A Potential Antiviral Strategy

Michel Liuzzi and Robert Déziel

Introduction

The herpes simplex viruses 1 and 2 (HSV-1 and HSV-2) represent two members of a group of at least seven human herpesviruses that includes varicella zoster virus (VZV), Epstein–Barr virus (EBV), cytomegalovirus (CMV), and the more recently identified human herpesviruses 6 and 7 (HHV-6 and HHV-7). Infections caused by HSV are widespread in the human population and serological surveys have indicated that at least 9 out of 10 individuals in the United States have antibodies against these viruses (Fields, 1990). HSV-1 causes oral-labial, facial, and ocular infections, as well as more severe diseases such as encephalitis, whereas HSV-2 is more generally associated with genital herpes infections.

The HSV genome is a double-stranded, linear DNA that encodes for more than 70 proteins, of which many are involved in nucleotide metabolism and DNA replication. The viral DNA is enclosed in a capsid which is surrounded by the envelope, a lipid bilayer, containing several essential glycoproteins. Lytic infections by HSV start by binding of the virus to the surface of the host cell, followed by fusion of the viral envelope with the plasma membrane and liberation of the nuclear capsid into the cytoplasm (Fields, 1990). The capsid is then degraded and the viral genome is released into the nucleus. Viral genes are transcribed by the cellular RNA polymerase II in a sequential and coordinated fashion. Regulatory proteins of the immediate early or α class are produced first and induce transcription of β or delayed early genes. The β mRNAs encode several proteins needed for viral DNA metabolism such as ribonucleotide reductase and DNA polymerase. Finally, the process of DNA replication stimulates late or γ gene

The Search for Antiviral Drugs
Julian Adams and Vincent J. Merluzzi, Editors
© 1993 Birkhäuser Boston

transcription and the resulting gene products, comprising mainly viral structural proteins which are required for assembly of new virus particles that may infect neighboring cells and initiate another infectious cycle.

In addition to lytic infections, HSV also have the capacity to establish latent infections in neurons, a hallmark of herpesviruses in general (Stevens, 1989). This type of infection is defined as the state in which the viral genome is present in neuronal cells without production of infectious virus. The latent virus can, however, be reactivated by ultraviolet (UV) light, stress, fever, and immunosuppression, thereby initiating a lytic infection in epithelial cells at the site of the initial infection.

Antiviral Strategy

Considering the large number of viral-specific processes that occur within a HSV-infected cell, it is anticipated that in-depth investigation of the reproductive cycle of the virus and its interaction with host cells may lead to the identification of a number of potential targets for chemotherapeutic intervention. By far, the most active area of research for the development of antiviral agents has been in the field of DNA synthesis inhibitors. Most of the compounds that have been approved for clinical use are nucleoside derivatives that act on the virally encoded DNA polymerase and terminate elongation of growing DNA chains. A challenging task in this area is to develop potent antiviral agents that will not interfere with the cellular DNA synthesis machinery. So far, one of the few compounds that has proven to be a selective antiherpetic agent is acyclovir (Reardon and Spector, 1991). Although acyclovir is now generally accepted as the drug of choice for the treatment of HSV infections, it has some limitations including induction of drug resistance (Freifeld and Ostrove, 1991) and limited efficacy against VZV and CMV. Undoubtedly, there is a need for more effective nucleoside analogues, but most importantly the increasing number of cases of resistance emphasizes the necessity to develop new antiviral drugs that exert their inhibitory activity by a different mechanism of action.

Since the discovery that HSV encodes for its own ribonucleotide reductase (RR), it has been suggested that the HSV RR may be essential for virus replication. Thus, it has been concluded that the HSV RR may represent an alternative antiviral target (Spector et al., 1985; Spector, 1989). Several reports have demonstrated that the viral ribonucleotide reductase is required for virus growth in resting cells (Goldstein and Weller, 1988; Jacobson et al., 1989) and for virulence of the virus in several animal studies (Cameron et al., 1988; Jacobson et al., 1989; Brandt et al., 1991; Yamada et al., 1991a; Idowu et al., 1992). Furthermore, recent studies have demonstrated that the HSV RR is also required for reactivation of latent infection (Jacobson et al., 1989). Taken together, these findings suggest that ribonucleotide reductase is essential for viral growth and therefore compounds that inhibit ribonucleotide reductase may turn out to be effective antiviral agents.

Properties of Ribonucleotide Reductases

Like the human and the *Escherichia coli* ribonucleotide reductases, the HSV RR catalyzes exclusively the reduction of ribonucleoside 5'-diphosphates to the corresponding deoxyribonucleotide counterparts (Figure 1) (Reichard, 1988). The HSV-encoded ribonucleotide reductase is a holoenzyme composed of two distinct subunits of the type $\alpha_2\beta_2$ (Figure 2). The large subunit (R1) is a dimer of two identical polypeptide chains with a molecular mass of 140 kD. The small subunit (R2) is also a homodimer of two 38-kD proteins. The R1 subunit contains redox-active cysteine residues, and the small subunit R2 contains a binuclear μ-oxo bridged iron center associated with a tyrosyl free radical. These chemical features of the viral ribonucleotide reductase are common to all known iron-dependent ribonucleotide reductases, and it is believed that they participate in the process of ribonucleotide reduction.

The molecular mechanism of enzymatic reduction of ribonucleotides has been derived from elegant studies conducted with recombinant *E. coli* ribonucleotide reductase (Stubbe et al., 1983; Erikson and Sjöberg, 1989). The catalytic site of the enzyme is thought to be located at the R1/R2 subunit interface. The R1 subunit binds the substrate and has redox-active sulfhydryl groups which provide the hydrogen for nucleotide reduction. The contribution of the small subunit R2 to the catalysis is to supply a radical. Initially, it had been proposed (Stubbe et al., 1983) that the tyrosyl radical of R2 directly abstracted the 3'-hydrogen of the nucleoside diphosphate substrate [see Figure 3]. However, the X-ray crystallographic structure of the *E. coli* R2 subunit (Nordlund et al., 1990) demonstrates that the tyrosyl radical is 10 Å away from the

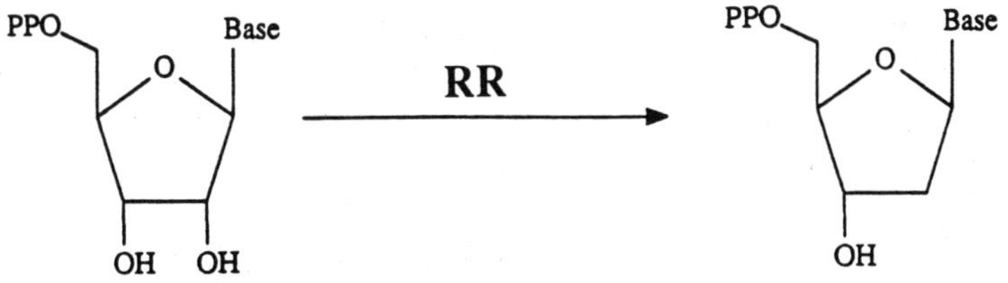

FIGURE 1. Reaction catalyzed by ribonucleotide reductase (RR).

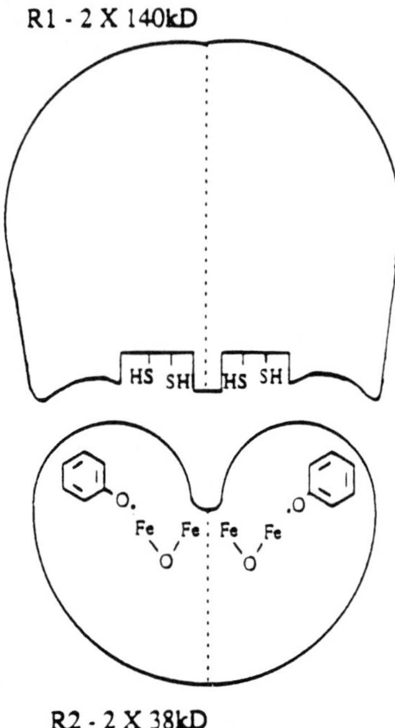

FIGURE 2. Schematic representation of the herpes simplex virus ribonucleotide reductase.

nearest surface of the enzyme and therefore presumably would not directly participate in hydrogen abstraction. This finding has led to the proposal that the sole function of the tyrosyl radical in R2 is to generate, by radical transfer, a protein radical on the large subunit in proximity to the substrate binding site (Stubbe, 1990; Bollinger et al., 1991). The R1 radical would then abstract the 3′-hydrogen atom of the substrate to yield the desired deoxyribonucleotide.

Inhibition of Ribonucleotide Reductase

The inhibition of ribonucleotide reductases has been under continued investigation in order to uncover novel antineoplastic and antiherpetic agents. Because of the intrinsic biochemical features discussed above, ribonucleotide reductases can be inhibited by a large variety of compounds including nucleoside analogues, metal chelators, and free radical scavengers (Spector, 1989). In this respect, thiosemicarbazone derivatives have attracted attention as potential agents to inhibit the viral ribonucleotide reductase. These compounds inactivate the HSV RR

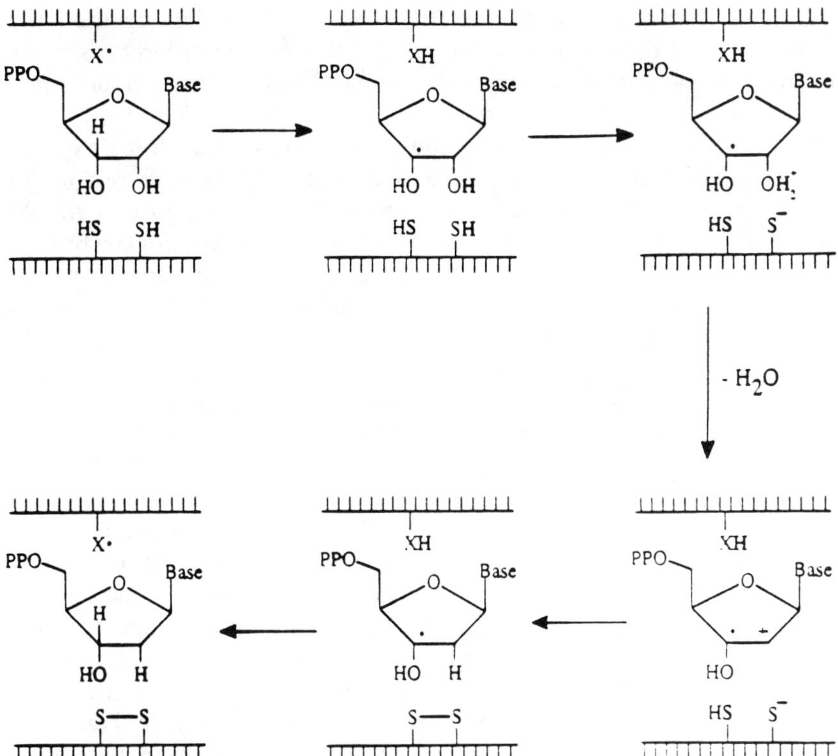

FIGURE 3. Proposed mechanism of ribonucleotide reduction. For details concerning the reaction mechanism, consult also Bollinger et al. (1991). X· represents the tyrosyl free radical on the small subunit. Adapted with permission of the American Society for Biochemistry and Molecular Biology from Stubbe J (1990): Ribonucleotide reductases: Amazing and confusing. *J Biol Chem* 265:5329–5332.

by disrupting the integrity of the free radical and/or iron center present in the small subunit of the enzyme. Since both the human and the viral small subunits contain these essential chemical features, it is not surprising that thiosemicarbazones and related compounds lack selectivity. Nonetheless, some compounds in this series appeared to (1) inhibit the HSV enzyme more efficiently than the human enzyme (Turk et al., 1986), (2) be effective in reducing the duration of cutaneous HSV-1 infections (Shipman et al., 1986), and (3) potentiate the antiherpetic activity of acyclovir (Spector et al., 1985; Lobe et al., 1991). Unfortunately, the major drawback of this class of compounds remains their toxicity in replicating cells. Thus, the potential utility of these ribonucleotide reductase inhibitors to serve as leads from which to develop orally active antiviral agents needs to be further investigated.

An alternative approach for the selective inhibition of the viral ribonucleotide reductase may derive from the marked differences in the primary amino acid sequences of the human and the viral ribonucleotide reductase subunits (Thelander and Thelander, 1989; Parker et al., 1991). This observation suggests that by identifying an essential domain on the viral ribonucleotide reductase that does not overlap with regions of sequence similarity with the human ribonucleotide reductase, it should be possible to selectively inactivate the HSV enzyme but not the human enzyme. In this regard, a serendipitous discovery by investigators from two independent laboratories has suggested a novel strategy for inhibiting the HSV ribonucleotide reductase (Cohen et al., 1986; Dutia et al., 1986). These investigators reported that the synthetic nonapeptide YAGAVVNDL (see Figure 5: HSV sequences for amino acid code) derived from the carboxy-terminal amino acid sequence of the HSV-1 ribonucleotide reductase small subunit, inhibited HSV-1 RR activity (IC_{50} = 38 μM) without affecting the mammalian ribonucleotide reductase. Since the inhibition was reversible and noncompetitive with respect to the substrate, it was hypothesized that the carboxy-terminal nine amino acids of the small subunit provided a binding domain for its association with the large subunit to generate active ribonucleotide reductase holoenzyme. The synthetic nonapeptide inhibited enzymatic activity by competing with the small subunit for binding to the large subunit, thereby preventing the formation of active enzyme (Figure 4). Taken together, these findings not only implied that the nonapeptide may represent a convenient lead from which to derive more effective HSV RR inhibitors, but they also generated renewed interest in the virally encoded enzyme as a potential target for antiviral therapy. Interestingly, it has been observed that ribonucleotide reductase small subunits from HSV-1 and HSV-2 share a high degree of sequence homology, at the carboxy-terminal end, with the ribonucleotide reductase small subunits of VZV, EBV, pseudorabies virus, and equine herpesvirus (Dutia et al., 1986; Davison and Scott, 1986; Baer et al., 1984; Cohen et al., 1990) (Figure 5). Therefore, it is tempting to speculate that HSV ribonucleotide reductase subunit association inhibitors may be effective against a wider spectrum of herpes viruses. Furthermore, subunit association via the carboxy terminus of the ribonucleotide reductase small subunits appears to be an essential feature of not only herpesviruses but also bacteria and mammals (Cosentino et al., 1991; Bushweller and Bartlett, 1991; Yang et al., 1990; Climent and Sjöberg, 1991). Since inhibition of ribonucleotide reductases by peptides corresponding to their respective small subunits has been shown to be universal and selective because of the inherent differences in the primary amino acid sequences (Cosentino et al., 1991), it may also be possible to design selective subunit association inhibitors of bacterial

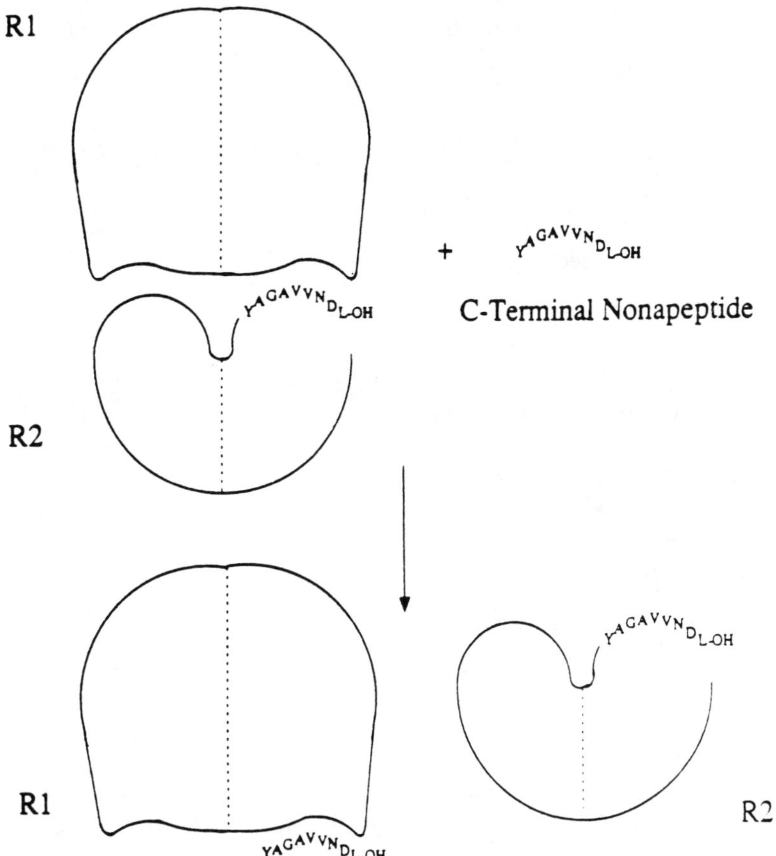

FIGURE 4. Proposed mechanism of ribonucleotide reductase inhibition by the C-terminal nonapeptide. *Note:* Recent evidence suggests that both C-terminal sequences of the R2 subunit may participate in the binding to R1. Therefore, R1 may actually contain two nonapeptide binding sites (see Climent and Sjöberg, 1991). Adapted with permission of Macmillan Magazines Limited from Dutia BM (1986): Specific inhibition of herpes ribonucleotide reductase by synthetic peptides. *Nature* 321:439–441.

and human ribonucleotide reductases that could result in a novel class of antibacterial and antineoplastic drugs.

Lead Optimization

Several investigators have reported that addition of the nonapeptide YAGAVVNDL to HSV-infected cells had no apparent effect on the yield of infectious virus (Dutia et al., 1986; Paradis et al., 1991). This lack of

Human   Asn-Ser-Phe-Thr-Leu-Asp-Ala-Asp-Phe-OH

HSV   Tyr-Ala-Gly-Ala-Val-Val-Asn-Asp-Leu-OH

VZV   Tyr-Ala-Gly-Thr-Val- Ile -Asn-Asp-Leu-OH

EBV   Tyr-Thr-Met-Leu-Val-Val-Asp-Asp-Leu-OH

FIGURE 5. C-terminal sequences of various ribonucleotide reductase small subunits.

antiviral effect was attributed to (1) low potency, (2) inability to enter cells, and (3) intrinsic susceptibility toward cellular proteases (Paradis et al., 1991; Telford et al., 1990). Accordingly, several laboratories have been interested in defining the structural requirements of an inhibitor which would have maximal inhibitory activity, proteolytic stability, and cell penetration capability. The ultimate goal would be to synthesize small inhibitors that would inhibit the HSV RR in vivo. In this regard, pioneering work by Langelier and co-workers has led to the disclosure of the peptide VVNDL as the minimum active segment of the carboxy-terminal nonapeptide (see Gaudreau et al., 1990). Furthermore, several structural features were identified which were useful to generate a series of ribonucleotide reductase inhibitors with potencies, in an enzyme activity assay, in the low micromolar range (Gaudreau et al., 1987, 1990; Paradis et al., 1991). For example, it was found that the carboxy-terminal nonapeptide derived from the VZV ribonucleotide reductase small subunit sequence (i.e., YAGTVIADL) was a better inhibitor of the HSV enzyme (IC_{50} = 13 μM) than the nonapeptide YAGAVVNDL (IC_{50} = 38 μM). Furthermore, N-terminal acetylation of the VZV nonapeptide yielded a more potent inhibitor (Ac-YAGTVIADL, IC_{50} = 2 μM) relative to its unacetylated analogue. Combined, these results suggested that it may be possible to further improve the potency of the peptidic inhibitors by identifying and optimizing specific molecular interactions between the inhibitors and the enzyme.

Mechanism of Peptide-Mediated Inhibition

A convincing piece of evidence in favor of the mechanism of action proposed above has been obtained by Paradis et al. (1988) by demonstrating that a photoaffinity analogue of the nonapeptide (4'-azido-Phe[6]-[3',5'-^{125}I]Tyr[7])-Ala-Gly-Ala-Val-Val-Asn-Asp-Leu with an IC_{50} of 80 μM was capable of specifically binding to HSV RR large subunit. To further substantiate that the interaction of ribonucleotide reductase association inhibitors with the large subunit was the primary cause of enzyme inhibition and also to determine the selectivity of binding to the viral large subunit, we used the photoaffinity probe 3-azido-4-hydroxy-[5-^{125}I]phenylpropionyl-(N-methyl-Val)-Ile-(Pyr-Asn)-Asp-Leu-OH (Figure 6), which belongs to a series of shorter inhibitors with an IC_{50} of 0.15 μM (Adams et al., 1991). Upon irradiation with 254-nm UV light the photoreactive probe cross-linked a 140-kD protein present in extracts isolated from HSV-1-infected cells (Lagacé et al., 1991). Photolabeling with the radioactive probe was specific since it could be competed with unlabeled analogue in a dose-dependent manner. No labeling was detected when mock-infected cell extracts were used. Thus, these findings not only corroborated that inhibitors based on the sequence of the carboxy terminus of the HSV small subunit inhibit enzymatic activity by binding specifically to the enzyme's large subunit (Paradis et al., 1988), but they also clearly indicated that the selectivity of binding was maintained with shorter compounds exhibiting increased inhibitory activity (Lagacé et al., 1991).

To investigate whether the nonapeptide YAGAVVNDL was indeed capable of interfering with subunit association, we performed immunoprecipitation experiments (Liuzzi et al., 1992) using the monoclonal antibody 535, which was raised against the small subunit of HSV-1 ribonucleotide reductase. This antibody was chosen since it was found to be nonneutralizing and to bind to the intact holoenzyme comprising both large and small subunits. This study demonstrated that in the absence of nonapeptide, the 535 antibody coimmunoprecipitated the small and the large subunits. In clear contrast however, when the holoenzyme was preincubated with the nonapeptide inhibitor, only the small subunit was detected by Western blotting, implying that the nonapeptide impeded the association of the holoenzyme's subunits (Liuzzi et al., 1992). These findings fully support the model proposed by others (Dutia et al., 1986; Cohen et al., 1986; McClements et al., 1988) that loss of enzymatic activity is directly related to separation of subunits.

The observation that the inhibited HSV RR exists as a mixture of

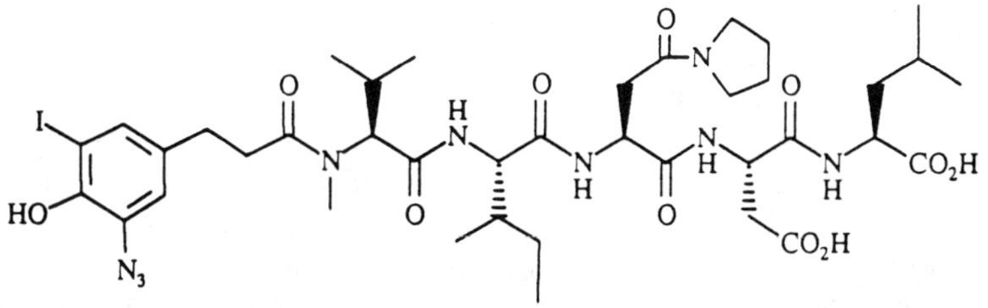

FIGURE 6. Structure of the photoaffinity probe 3-azido-4-hydroxy-[5-^{125}I]-phenylpropionyl-(N-methyl-Val)-Ile-(Pyr-Asn)-Asp-Leu-OH.

dissociated subunits raised the concern that viral subunits may interact and associate with mammalian subunits to form a hybrid ribonucleotide reductase. Depending on the precise nature of this putative association, it is possible that the hybrid reductase activity might be insensitive to ribonucleotide reductase association inhibitors that are directed against the viral enzyme. To test this possibility we have cloned and expressed recombinant HSV and mammalian large and small subunits, and then used them in reconstitution experiments in order to evaluate the selectivity of subunit interactions (Liuzzi et al., 1991). Neither subunit had reductase activity when tested alone. Fully functional enzyme was only obtained when recombinant viral small subunit, but not mammalian small subunit, was combined with HSV large subunit. The reconstituted reductase activity was specifically inhibited by the N-terminally acetylated nonapeptide Ac-YAGTVIADL derived from the carboxy terminus of the VZV RR small subunit. It was also observed that the inhibition of the viral enzyme was reversed by adding to the reaction mixtures either the excess HSV large subunit or the excess HSV small subunit but not the mammalian subunits. Thus, these results indicated that the carboxy-terminal peptides and the HSV small subunit shared a common binding site on the HSV large subunit. Since the mammalian subunits did not compete with the peptide inhibitor for binding to the viral large subunit, our findings also suggested that the formation of an active hybrid ribonucleotide reductase holoenzyme is unlikely.

Conclusion

The experimental evidence presented here corroborates previous reports by others showing that peptides based on the sequence of the carboxy terminus of the small subunit of HSV RR inhibit enzymatic activity by binding specifically and selectively to the ribonucleotide reductase large subunit. Thus, inhibition of ribonucleotide reductase by subunit association inhibitors may prove to be a valid strategy to interfere with viral reproduction for the treatment of infections caused by herpesviruses. It remains to be established, however, whether inhibition of ribonucleotide reductase by subunit association inhibitors can or cannot lead to inhibition of HSV replication. The results presented here certainly indicate that the potency of HSV ribonucleotide reductase inhibitors can be significantly improved without affecting the mechanism of action and the specificity of the inhibitors. Moreover, the recent observation that ribonucleotide reductase inhibitors potentiate the action of acyclovir (Spector et al., 1985; Lobe et al., 1991) and that ribonucleotide reductase mutant viruses appear to be hypersensitive to the antiviral effect of acyclovir (Coen et al., 1989; Yamada et al., 1991b) add supporting evidence to the proposition that the HSV RR is essential for viral replication. Thus, it is anticipated that ribonucleotide reductase subunit association inhibitors should be effective to inhibit HSV replication.

Acknowledgments

We wish to thank N. Moss for artwork and for critical review of the manuscript, L. Thelander and R. Ingemarson for procuring antibodies 535 and 932, L. Lagacé and J. C. Chafouleas for helpful comments on the manuscript, and R. Plante and J. Gauthier for synthesizing the inhibitory peptides.

References

Adams J, Beaulieu PL, Déziel R, DiMaio J, Grenier L, Lavallée P, Moss N (1991): Antiherpes pentapeptide derivatives having a substituted aspartic acid side chain. Eur Pat App 411,334

Baer R, Bankier AT, Biggin MD, Deininger PL, Farewell PJ, Gibson TJ, Hatfull G, Hudson GS, Satchwell SC, Séguin C, Tuffnell PS, Barrell BG (1984): DNA sequence and expression of the 895-8 Epstein–Barr virus genome. *Nature (London)* 310:207–211

Bollinger JM, Edmondson DE, Huynh BH, Filley J, Norton JR, Stubbe J (1991): Mechanism of assembly of the tyrosyl radical-dinuclear iron cluster cofactor of ribonucleotide reductase. *Science* 253:292–298

Brandt CR, Kintner RT, Visalli RJ, Pumfrey AM, Grau DR (1991): The herpes simplex virus ribonucleotide reductase is required for ocular virulence. *J Gen Virol* 72:2043–2049

Bushweller JH, Bartlett PA (1991): Investigation of an octapeptide inhibitor of *Escherichia coli* ribonucleotide reductase by transferred nuclear Overhauser effect spectroscopy. *Biochemistry* 30:8144–8151

Cameron JM, McDougall I, Marsden HS, Preston VG, Ryan MD, Subak-Sharpe JH (1988): Ribonucleotide reductase encoded by herpes simplex virus is a determinant of the pathogenicity of the virus in mice and a valid antiviral target. *J Gen Virol* 67:2607–2612

Climent I, Sjöberg B-M (1991): Carboxyl-terminal peptides as probes for *Escherichia coli* ribonucleotide reductase subunit interaction: Kinetic analysis of inhibition studies. *Biochemistry* 30:5164–5171

Coen DM, Goldstein DJ, Weller SK (1989): Herpes simplex virus ribonucleotide reductase mutants are hypersensitive to acyclovir. *Antimicrob Agents Chemother* 33:1395–1399

Cohen EA, Gaudreau P, Brazeau P, Langelier Y (1986): Specific inhibition of herpes virus ribonucleotide reductase by a nonapeptide derived from the carboxy terminus of subunit 2. *Nature (London)* 321:441–443

Cohen EA, Paradis H, Gaudreau P, Brazeau P, Langelier Y (1990): Identification of viral polypeptides involved in pseudorabies virus ribonucleotide reductase activity. *J Virol* 61:2046–2049

Cosentino G, Lavallée P, Rakhit S, Plante R, Gaudette Y, Lawetz C, Whitehead PW, Duceppe J-S, Lépine-Frenette C, Dansereau N, Guilbeault C, Langelier Y, Gaudreau P, Thelander L, Guindon Y (1991): Specific inhibition of ribonucleotide reductases by peptides corresponding to the C-terminal of their second subunit. *Biochem Cell Biol* 69:79–83

Davison AJ, Scott JE (1986): The complete DNA sequence of varicella-zoster virus. *J Gen Virol* 67:1759–1816

Dutia BM, Frame MC, Subak-Sharpe JH, Clark WN, Marsden HS (1986): Specific inhibition of herpes ribonucleotide reductase by synthetic peptides. *Nature (London)* 321:439–441

Erikson S, Sjöberg B-M (1989): Ribonucleotide reductase. In: *Allosteric Enzymes*, Hervé G, ed. Boca Raton, FL: CRC Press, pp 189–215

Fields, BN (1990): *Virology*. New York: Raven Press.

Freifeld AG, Ostrove JM (1991): Resistance of viruses to antiviral drugs. *Annu Rev Med* 42:247–259

Gaudreau P, Michaud J, Cohen EA, Langelier Y, Brazeau P (1987): Structure–activity studies on synthetic peptides inhibiting herpes simplex virus ribonucleotide reductase. *J Biol Chem* 262:12413–12416

Gaudreau P, Paradis H, Langelier Y, Brazeau P (1990): Synthesis and inhibitory potency of peptides corresponding to the subunit 2 C-terminal region of herpes virus ribonucleotide reductases. *J Med Chem* 33:723–730

Goldstein DJ, Weller SK (1988): Herpes simplex virus type 1–induced ribonucleotide reductase activity is dispensable for virus growth and DNA synthesis: Isolation and characterization of an ICP6 lac Z insertion mutant. *J Virol* 62:196–205

Idowu AD, Fraser-Smith EB, Poffenberger KL, Herman RC (1992): Deletion of the herpes simplex virus type 1 ribonucleotide reductase gene alters virulence and latency *in vivo*. *Antiviral Res* 17:145–156

Jacobson JG, Leib DA, Goldstein DJ, Bogard CL, Schaffer PA, Weller SK, Coen DM (1989): A herpes simplex virus ribonucleotide reductase deletion mutant is defective for productive acute and reactivatable latent infections of mice and for replication in mouse cells. *Virology* 173:276–283

Lagacé L, Liuzzi M, Thibeault D, Dô F, Chafouleas JG, Gauthier J, Plante R, Déziel R, Guindon Y (1991): Specificity of binding of HSV ribonucleotide reductase inhibitors with the large subunit of the enzyme. *Program 16th Int Herpes Virus Workshop, Pacific Grove, California*, p 66

Liuzzi M, Lapierre F, Ingemarson R, Lagacé L, Massie B, Fleurent J, Thelander L, Chafouleas JG (1991): Evaluation of subunit association utilizing recombinant R1 and R2 of HSV ribonucleotide reductase. *Program 16th Int Herpes Virus Workshop, Pacific Grove, California*, p 72

Liuzzi M, Scouten E, Ingemarson R (1992): Inhibition of herpes simplex ribonucleotide reductase by synthetic nonpeptides: A potential antiviral therapy. In: *Innovations in Antiviral Development and the Detection of Virus Infections*, Walsh L, Block TM, Crowell R, Jungkind DL, eds., New York: Plenum Press, pp 129–138

Lobe DC, Spector T, Ellis MN (1991): Synergistic topical therapy by acyclovir and A1110U for herpes simplex virus induced zosteriform rash in mice. *Antiviral Res* 15:87–100

McClements W, Yamamaka G, Garsky V, Perry H, Bacchetti S, Colonno R, Stein RB (1988): Oligopeptides inhibit the ribonucleotide reductase of herpes simplex virus by causing subunit separation. *Virology* 162:270–273

Nordlund PI, Sjöberg B-M, Eklund H (1990): Three-dimensional structure of the free radical protein of ribonucleotide reductase. *Nature (London)* 345:593–598

Paradis H, Gaudreau P, Brazeau P, Langelier Y (1988): Mechanism of inhibition of herpes simplex virus (HSV) ribonucleotide reductase by a nonapeptide corresponding to the carboxy terminus of its subunit 2: Specific binding of a photoaffinity analog, [4′-azido-Phe^6] HSV H$_2$-(6–15), to subunit 1. *J Biol Chem* 263:16045–16050

Paradis H, Langelier Y, Michaud J, Brazeau P, Gaudreau P (1991): Studies on *in vitro* proteolytic sensitivity of peptides inhibiting herpes simplex virus ribonucleotide reductase lead to discovery of a stable and potent inhibitor. *Int J Pept Protein Res* 37:72–79

Parker NJ, Begley CG, Fox RM (1991): Human M1 subunit of ribonucleotide reductase: cDNA sequence and expression in stimulated lymphocytes. *Nucleic Acids Res* 19:3741

Reardon JE Spector T (1991): Acyclovir: Mechanism of antiviral action and potentiation by ribonucleotide reductase inhibitors. *Adv Pharmacol* 22:1–27

Reichard P (1988): Interactions between deoxyribonucleotide and DNA synthesis. *Annu Rev Biochem* 57:349–374

Shipman C Jr, Smith SH, Drach JC, Klayman DL (1986): Thiosemicarbazones of 2-acetylpyridine, 2-acetylquinoline, 1-acetylquinoline, and related compounds as inhibitors of herpes simplex virus *in vitro* and in a cutaneous herpes guinea pig model. *Antiviral Res* 6:197–222

Spector T (1989): Ribonucleotide reductase encoded by herpes viruses: Inhibitors and chemotherapeutic considerations. In: *International Encyclopedia of Pharmacology and Therapeutics*, Cory JG, Cory AH, eds. Elmsford, NY: Pergamon Press, pp 235–243

Spector T, Averett DR, Nelson DJ, Lambe CV, Morrison RW, St Clair MH, Furman PA (1985): Potentiation of antiherpetic activity of acyclovir by ribonucleotide reductase inhibition. *Proc Natl Acad Sci USA* 82:4254–4257

Stevens JG (1989): Human herpes viruses: A consideration of the latent state. *Microbiol Rev* 53:318–332

Stubbe J (1990): Ribonucleotide reductases: Amazing and confusing. *J Biol Chem* 265:5329–5332

Stubbe J, Ator M, Krenitsky T (1983): Mechanism of ribonucleotide diphosphate reductase from *Escherichia coli*. Evidence for 3′-C−H bond cleavage. *J Biol Chem* 258:1625–1630

Telford E, Owsianka A, Marsden HS (1990): Stability of the herpes virus ribonucleotide reductase-inhibiting nonapeptide YAGAVVNDL in extracts of HSV 1–infected cells. *Antiviral Chem Chemother* 1:223–226

Thelander M, Thelander L (1989): Molecular cloning and expression of the functional gene encoding the M2 subunit of mouse ribonucleotide reductase: A new dominant marker gene. *EMBO J* 8:2475–2479

Turk SR, Shipman C Jr, Drach JC (1986): Selective inhibition of herpes simplex virus ribonucleotide diphosphate reductase by derivatives of 2-acetylpyridine thiosemicarbazone. *Biochem Pharmacol* 35:1539–1545

Yamada Y, Kimura H, Morishima T, Daikoku T, Maeno K, Nishiyama Y (1991a): The pathogenicity of ribonucleotide reductase null mutants of herpes simplex virus type 1 in mice. *J Infect Dis* 164:1091–1097

Yamada Y, Yamamoto N, Daikoku T, Nishiyama Y (1991b): Susceptibility of a herpes simplex virus ribonucleotide reductase null mutant to deoxyribonucleosides and antiviral nucleoside analogs. *Microbiol Immunol* 35:681–686

Yang FD, Spanevello RA, Celiker I, Hirschmann R, Rubin H, Cooperman BS (1990): The carboxyl terminus heptapeptide of the R2 subunit of mammalian ribonucleotide reductase inhibits enzyme activity and can be used to purify the R1 subunit. *FEBS Lett* 272:61–64

Keyword Index

This index was established according to the keywords supplied by the authors. Page numbers refer to the beginning of the chapter.